Ju 87 Stuka

in action

By Brian Filley

Color by Don Greer

Illustrated by James G. Robinson

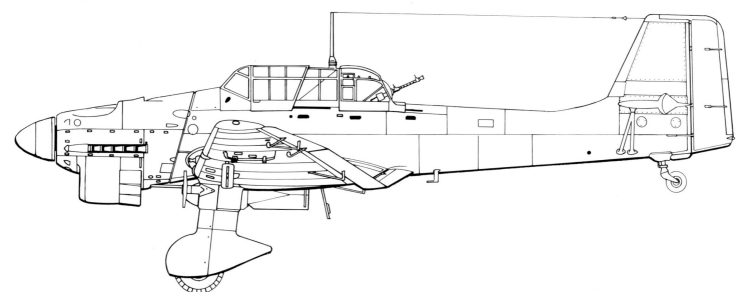

Aircraft Number 73

squadron/signal publications

Ju 87R-2/Trops of 4./Stukageschwader 2 on patrol against British positions in North Africa during autumn of 1941. Like many early desert campaign Stukas, these aircraft have had Sand Yellow applied over their upper surface European splinter camouflage. Stukas would later be delivered to North Africa in official factory applied desert colors.

Junkers Factory Emblem

ISBN 0-89747-175-X

If you have any photographs of the aircraft, armor, soldiers or ships of any nation, particularly wartime snapshots, why not share them with us and help make Squadron/Signal's books all the more interesting and complete in the future. Any photograph sent to us will be copied and the original returned. The donor will be fully credited for any photos used. Please indicate if you wish us not to return the photos. Please send them to: Squadron/Signal Publications, Inc., 1115 Crowley Dr., Carrollton, TX 75011-5010.

ACKNOWLEDGEMENTS

Special thanks go to Ernie Vagi for a peek into his photo collection and personal library; to Anita Faber and Helga Gerhardt for their kind translations of German text and remembrances of bygone times; and to Larry Davis and Jim Mesko for their encouragement.

PHOTO CREDITS

Bundesarchiv
E C P Armee
Hans Obert
Ernest Vagi
Hans Redemann
Imperial War Museum
Richard F Grant
Squadron Signal Archives
Smithsonian Institution
U S Air Force
Zdenek Titz

(Right) Ju 87B-2s of 7./Stukageschwader 77 being prepared for action on the bitterly contested Russian Front. Ultimately, every production variant of the Stuka, except the 'Anton', would be committed to the 'struggle in the East', that lasted from 22 June 1941 until May of 1945. (Bundesarchiv)

3

INTRODUCTION

Few aircraft in World War II earned their place in aviation history with as much pure infamy as Germany's Junkers Ju 87 dive bomber. It was known simply as the *Stuka*, a contraction of the German word *Sturzkampfflugzeug* (diving battle aircraft) which was the generic German term for all dive-bombers, but became synonymous with the Junkers Ju 87 by both the Allies and the Axis. It was a slow and angularly ugly aircraft that was easily victimized by opposing fighters, and yet an aircraft which wrought more destruction and outright terror over a greater period of time than any other aircraft of its type. In the process *Stuka* became a household word on both sides, and its dual-identity as both a *super terror weapon* and an *embarrassing failure*, created a legend. Whatever its drawbacks, the *Stuka* was regarded as a total success by the men who designed it, the strategists who had thrust it into battle, and the countless *Wehrmacht* troops who depended on its protective 'flying artillery'. 'Where the infantry went, so went the *Stuka*', and this was the purpose for which the Ju 87 had been created.

During the First World War, Germany had utilized squadrons of trench-strafing attack aircraft with great success, and several manufacturers were, at one time or another, encouraged to produce specialized aircraft to meet this revolutionary new combat role. Of these, the Junkers Flugzeugwerke A G of Dessau, Germany — founded by the enterprising Professor Hugo Junkers — was responsible for some of the world's first all-metal stressed-skin designs, and quickly became an established leader in the field. With the end of the war, and the subsequent armament restrictions of the Versailles Treaty, Hugo Junkers directed his firm to concentrate on the production of civil airliners and sport-planes, but with a continuing eye toward forbidden German warplane development. Junkers established subsidiary firms in Turkey, Russia and at Malmo-Linhamm, Sweden (A B Flygindustri) to legally escape the Allied restrictions. The majority of this Junkers 'outside' research was devoted to metal-skinned ground-attack planes and light bombers, which were exported to the air forces of other nations.

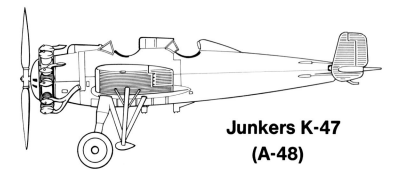

Junkers K-47
(A-48)

Junkers' first direct exposure to dive-bombing, which was then gaining international recognition, came with the first flight of the Junkers K-47 monoplane at the Swedish works in 1928. Although the two-seat K-47 proved to be lackluster in its intended fighter/interceptor role, its sturdily braced wings offered an ideal test-bed for diving experiments. With the support of Swedish military authorities, diving trials were shortly being conducted by a Junkers-Dessau research team in Sweden, whose findings, it was hoped, would form the basis of a new and up-to-date dive-bomber design. Extensive tests with Swedish-produced K-47s and Dessau-produced A-48s would span several years, and help to develop efficient bomb racks, warload ballistics, diving techniques and diving sights. K-47s and A-48s continued to be at the nucleus of more official German dive-bombing research after Hitler came to power in 1933, assisting the preparation of a dive-bomber contingent within the new as yet clandestine *LUFtwaffe*.

1936 Sturzbomber Competitors

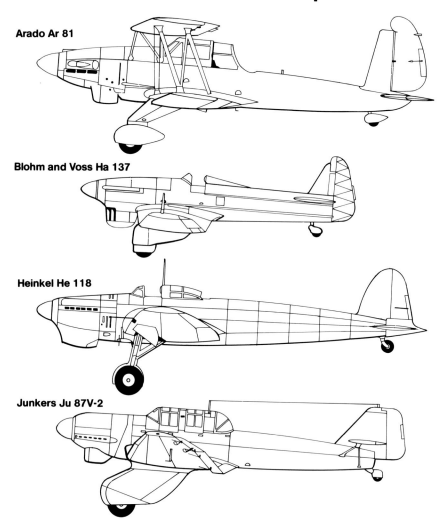

Arado Ar 81

Blohm and Voss Ha 137

Heinkel He 118

Junkers Ju 87V-2

Even before the establishment of the Third Reich however, the colorful WWI ace and sports flyer, Ernst Udet, had been campaigning for a German dive-bomber after being inspired by US Navy diving demonstrations at air shows in the United States. Under Udet's constant urging, and with regard to the possible need for such an aircraft to serve as army-support, when the Nazis came to power the planners of the newly-formed *Reichsluftfahrtministerium* (RLM) the State Aviation Ministry launched a two-phase *Sturzbomber* development program in 1933. Unfortunately a deep seated controversy soon developed among RLM officials over the practicality of 'suicidally' diving an aircraft against a well-defended enemy target. This controversy would become a bitterly debated issue which would almost lead to the cancellation of the entire *Sturzbomber Programm* three years later. Nevertheless, the RLM *Technischen Amt* (Technical Office) released specifications for the second-phase 'modern dive-bomber' design competition in January of 1935 — by which time the more conventional Hs 123 biplanes of the 'first phase' or *Sofort Programm* (Immediate Program) were entering their final development and would, within the next two years, establish the first operational *Stukagruppen*.

(Above) The first prototype Ju 87V-1 with the early shallow radiator bath and original twin tail assembly inherited from the Junkers K-47. (Zdenek Titz)

Ju 87V-1 (Kestrel Engine)

 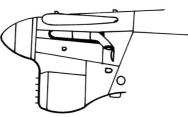

Original Radiator   **Enlarged Radiator**

As these events were unfolding the ageing Hugo Junkers had been forced to retire from his company the previous year, because of his political beliefs. With direct control of the Junkers-Flugzeugwerke being assumed by the state, greater priority was given to the development of Junkers' military aircraft and aero-engine production. The strides which Junkers had made in dive-bombing research had greatly influenced the new RLM specifications, which meant that the specifications were virtually tailor-made for the Junkers' designs entered in the *Sturzbomber* competition. In fact, advanced approval had already been granted for just such a project, and work had already begun on a Junkers dive-bomber, the Ju 87.

Drafting work on the Ju 87 had been initiated during the autumn of 1933 under the direction of Dipl.-Ing. Hermann Pohlmann, who himself had been a bomber pilot during the 'Great War'. Drawing from his own experience with the development of the K-47, Pohlmann patterned his twin-tailed dive-bomber along the straight-forward 'no-frills' design principals of the period. By late 1934 a wooden mockup of the Ju 87 had passed inspection by the RLM, with actual construction to take place over the next year. The thick two-spar wings were an inverted-gull configuration to provide optimum diving strength, clearance for both the propeller and bomb load, and allow for the shortest possible under-carriage to reduce drag. A hefty trestle-like fixed landing gear eliminated the need for a load-restrictive retraction mechanism, and was to be additionally braced to the oval-section fuselage with aerofoil-section struts. The fuselage was assembled in upper and lower halves and featured an enclosed glazed cockpit with swinging hatches for crew entry. A thick frame in the center of the greenhouse protected the crew in the event that the aircraft overturned. The fuselage structure was completely

(Above) The Ju 87V-4 in its original form, with smooth cowling side panels and before the rearmost canopy windows have been installed. (Zdenek Titz)

Underwing Dive Air-Brakes

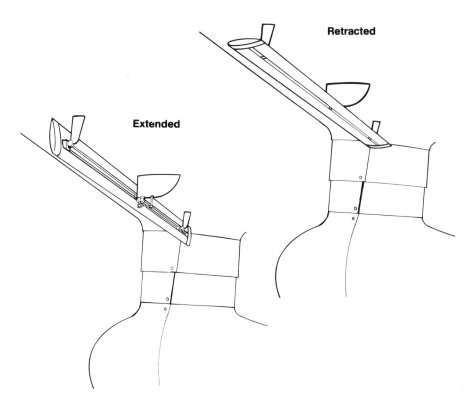

Retracted

Extended

enclosed in tough all-metal duraluminum stressed-skinning for maximum strength. Due to the lack of an efficient German-made powerplant, the first prototype was powered by a fully supercharged British liquid cooled Vee 12-cylinder 525 hp Rolls-Royce Kestrel V driving a fixed pitch two-bladed wooden propeller.

On 17 September 1935, the Ju 87V-1 took to the air from the test field at Dessau with only minor problems in stability. Despite its ungainly appearance, the new dive-bomber displayed relatively smooth handling qualities; the unique Junkers' floating flaps and ailerons contributing to its lightness of control. The only required modification was an enlargement of the chin radiator to counter persistent overheating problems with the Kestrel engine. In this form, the modified Ju 87V-1 continued its flight testing until 24 January 1936, when it suddenly crashed during a test dive, killing Junkers' chief test pilot, Willy Neuenhofen, and his observer — which was partially caused by a lack of diving airbrakes which had not yet been installed beneath the outer wings. A subsequent investigation focused additional suspicions on the twin vertical tail assembly, and these were replaced by a single central fin configuration on the next two prototypes.

Beginning with the Ju 87V-2 (D-UHUH), a totally reworked cruciform empennage was introduced, with a broad vertical tail and round-tipped horizontal surfaces. By now a suitable German powerplant was available, a 610 hp inverted-Vee Junkers Jumo 210 Aa, whose installation permitted a redesign of the cowling, eliminating the upper fairings which had been necessary with the upright-Vee Kestrel.

The Ju 87V-3 (D-UKYQ), the third prototype differed from the Ju 87V-2 only in having a still larger vertical fin and rudder, squared-off horizontal tailplanes and lowered engine mounts to create a downward slant to the cowling, improving the pilot's forward vision. Both the second and third prototypes were fitted with three-bladed Jumo-Hamilton metal airscrews and slat-like dive-brakes under the outboard wing panels. These simple, aerofoil-shaped units pivoted flatly into the airstream during a dive to safely control speed.

That March, with the Ju 87V-3 in reserve, the Ju 87V-2 was transferred to the *Erprobungsstelle* (Experimental Test Center) at Rechlin to begin qualifying trials against three additional contenders for the RLM contract. The Arado 81 biplane and the single-seat Blohm und Voss Ha 137 were rejected out of hand since neither fulfilled the concept of a heavy twin-seat monoplane. However, the sleeker Heinkel He 118 with a retractable undercarriage presented a more serious threat. The less sophisticated Ju 87 could be plummeted into accurate near-vertical dives, and its simpler structure would assure greater ease of production and field maintenance. In contrast, the He 118 averaged a diving angle of only 50 degrees, field-servicing would be more complicated, and its overall design more closely resembled a light attack-bomber than a true dive-bomber. With favoritism already leaning the RLM toward the Ju 87, the non-fatal crash of the He 118V-3 on 27 June 1936, destroyed any last chance for the Heinkel entry (especially since the pilot on that occasion happened to be the new head of the RLM *Technische Amt*, Ernst Udet). A production contract was awarded to the Junkers Flugzeugwerke.

In November the final prototype, the Ju 87V-4 (D-UBIP), was introduced. Further refinements found on this 'productionized' version included squared-off vertical tail fin, and a blending of the aft canopy glazings into the upper decking of the rear fuselage. The engine mounts were again lowered for greater forward visibility, and a small viewing window was installed beneath the forward fuselage enabling the pilot to sight his target prior to beginning his diving attack. Aiming was accomplished with a *Stukavisier* or *Stuvi* (diving-sight) as it was called, mounted above the instrument panel. Later a series of angled lines were painted on the inside of the starboard cockpit window to allow the pilot to judge his diving angle with the horizon. In order to clear the dropping bomb away from the propeller arc, a forward-swinging rack was added beneath the fuselage, hinging at the firewall point. Offensive armament was augmented with a forward firing 7.92MM Rheinmetall-Borsig MG 17 machine gun mounted in the starboard wing, and rear defensive armament was provided by a a flexible 7.92MM MG 15 in the observers' position.

Another innovation, which had originally been explored with the earlier K-47, was an automatic diving pull-out device, which coordinated the underwing dive-brakes, elevator trim and bomb-release mechanism to safely recover the aircraft in case of pilot 'black-out' brought on by g-stress. The automatic pull-out device was initiated at the outset of a dive with the extension of the dive-brakes and was activated upon release of the bombs. This system was later augmented by an automatic altimeter setting at 1,475 feet, which could be manually over-ridden by the pilot if necessary, by exerting a strong pull on the cockpit control stick.

Final Prototype Changes

Jumo 210 Aa Engine (Early V-4)

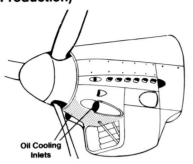

Jumo 210 Ca Engine (Production)

Oil Cooling Inlets

Wings

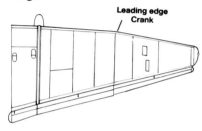

Leading edge Crank

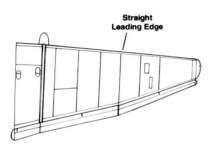

Straight Leading Edge

Extensive bombing tests were conducted with the Ju 87V-4 through the spring of 1937, using a variety of fuses and payloads. However, the heavy, underpowered Ju 87 could only lift its maximum bombload of 1,102 pounds if the second crew member was eliminated as weight compensation. Prior to production several further improvements were introduced, some of which were retro-fitted to the Ju 87V-4 during trials. For an extra margin of lifting power, the 610 hp Jumo Aa was replaced by a 640 hp Jumo 210 Ca engine, and the oil cooling system was improved with enlarged twin inlets beneath the forward nose and outlet vents on the cowling side-panels below the exhausts. To ease assembly the 'crank' in the outer wing leading edge was brought to a straight edge, with wing area being maintained by increasing the taper of the trailing edge. The rudder profile was further squared by eliminating the lower-edge droop. With these final revisions the assembly of the first pre-production series was commenced under the designation Ju 87A-0. The A series would be known as *Anton* according to the German radio-phonetics system. The *Luftwaffe* now had its first modern dive-bomber and the only remaining obstacle was a persistant anti-dive bomber sentiment within the divided Reichsluftfahrt-ministerium.

(Below) Seen at a later period in its testing, the Ju 87V-4 has been fitted with production cooling scoops to the lower cowling and full armament. The gunners' MG 15 aperature was little more than a large vertical slot. The droop to the lower rudder of the V-4 was straightened in its final configuration.

Ju 87 DEVELOPMENT

Ju 87A-1

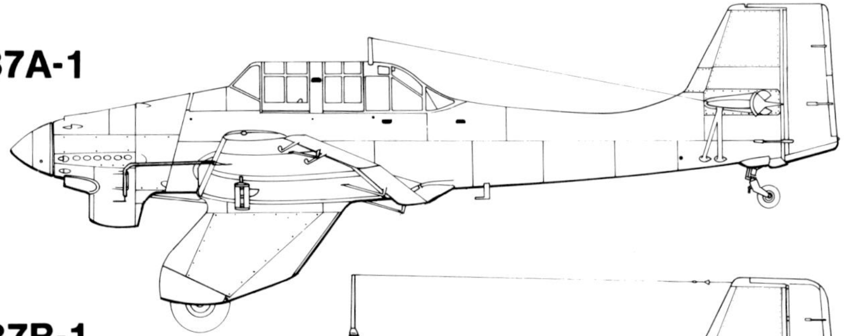

Ju 87B-1

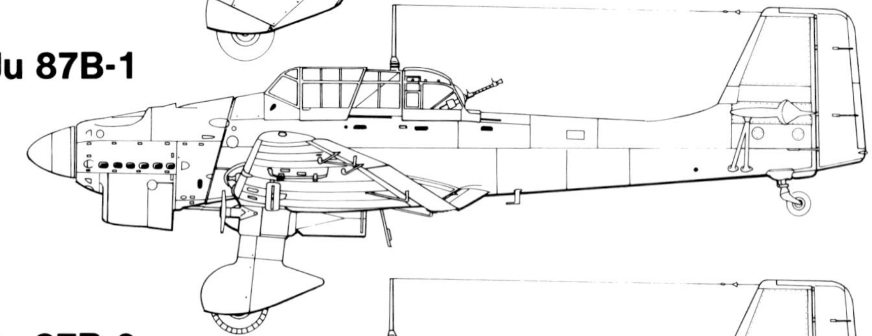

Ju 87B-2

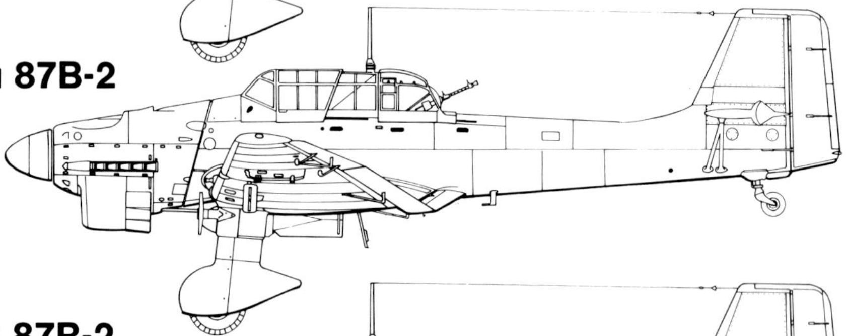

Ju 87R-2

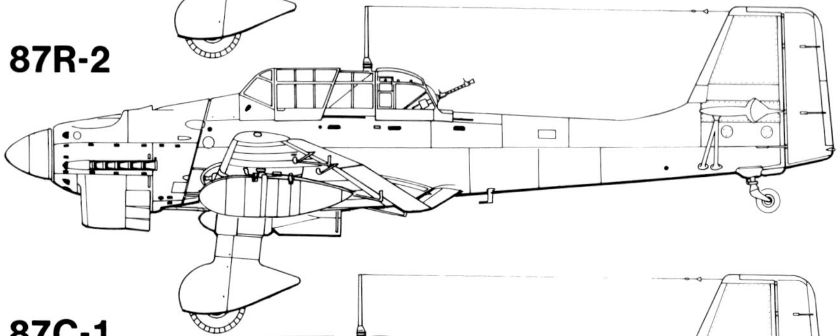

Ju 87C-1

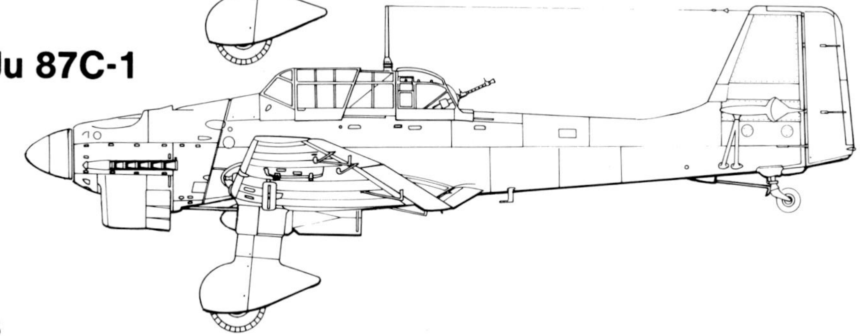

Ju 87D-1 (Early)

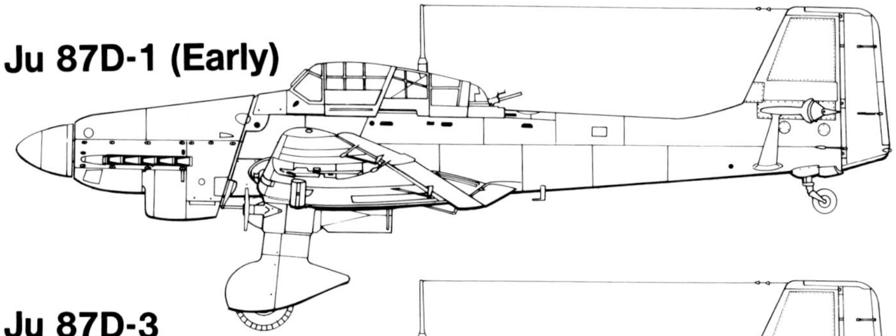

Ju 87D-3

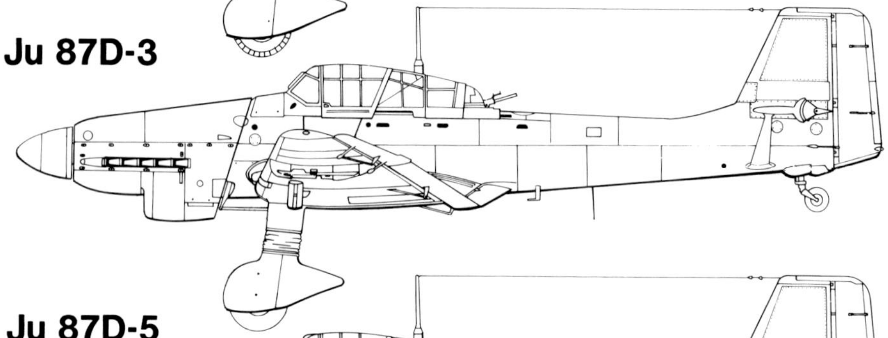

Ju 87D-5

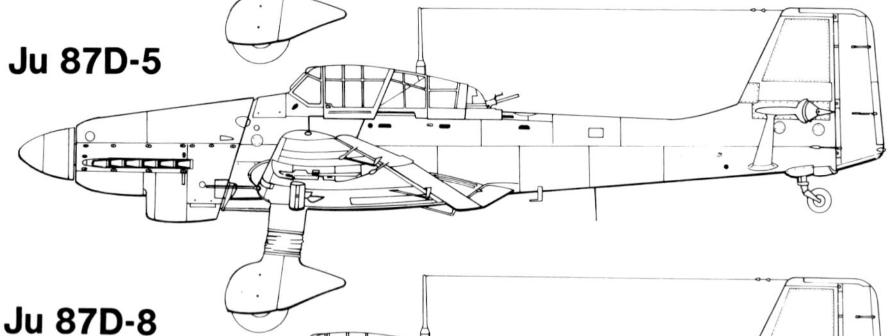

Ju 87D-8

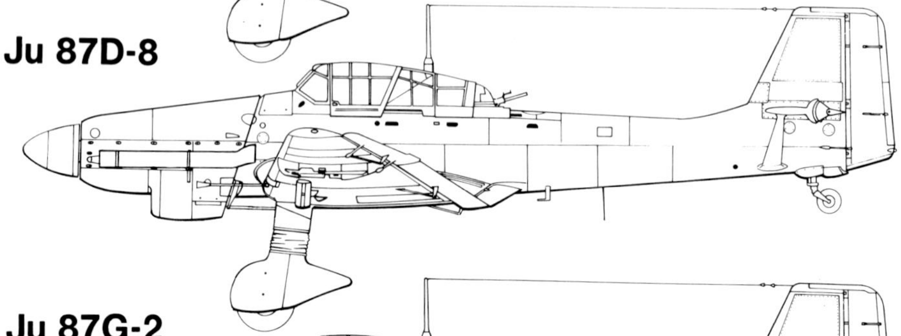

Ju 87G-2

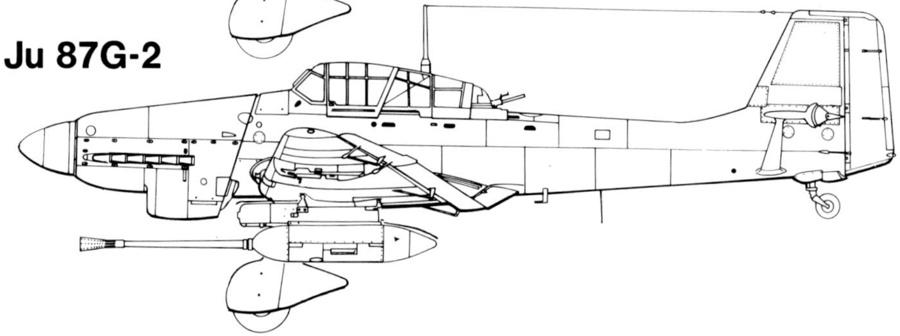

Ju 87A

Despite the continued grumbling of dive-bomber critics, the first of the ten pre-production Ju 87A-Os were rolled off the Dessau assembly line in early 1937. These initial *Antons* had only an average performance by contemporary dive-bombing standards, with an unloaded top speed of only 199 mph, a maximum diving speed of 279 mph and an unloaded range of 620 miles. Normal offensive load was still restricted to a single 250 kg (551 pound) bomb since the maximum 1,102 pound bomb load could still only be carried when the aircraft was flown without the gunner.

Following factory trials, the ten pre-production Ju 87A-Os were assigned to I/*Stukageschwader 162 'Immelmann'* for service orientation. Slowly but steadily deliveries of production Ju 87A-1s, identical to the Ju 87A-0, began replacing the biplanes of the *Stukagruppen* by mid-year. Germany's involvement in the Spanish Civil War provided a perfect testing ground for the new weapon, and three *Immelmann Geschwader* Ju 87A-1s were delivered to *Kampfgruppe K.88* of the *Luftwaffe's Legion Condor* for service evaluation under combat conditions. After some early teething difficulties, the new dive-bombers were flown in combat for the first time in February of 1938, piloted exclusively by German crews on a rotation basis to provide the maximum number of personnel with combat experience. Their success in several key offensives erased all doubt of the Stuka's possibilities, and the first coordinated tactics between precision Stuka attacks and the movement of ground forces began to emerge. The shape of the future was being molded.

Ju 87A-2

By late 1937, manufacture of the Ju 87A-1 had been superseded by the modestly improved Ju 87A-2, featuring an uprated 680 hp Jumo 210 Da engine with a two-stage supercharger and improved communications equipment. Later Ju 87A-2s had a further rounding-off of the upper rudder. The Jumo 210 Da powered Ju 87A-2 produced only a nominal increase in performance and fell far short of solving *Anton's* persistent load-lifting problem. From the pilot's viewpoint, the bothersome manual operation of the propeller pitch, supercharger and radiator shutters (both before and after a dive) was a further unpopular characteristic which compounded *Anton's* sluggishness with unnecessary difficulty. Fortunately, a successor to the Ju 87A series was already under development at Dessau, and as a result *Anton* production was terminated at the end of 1938 after a total of 262 aircraft were built. Of these, 192 had been produced at the main Dessau plant, with the balance coming off the new Junkers' Weser assembly line at the Berlin-Tempelhof airport.

Other than the action seen by the trio of Ju 87A-1s in Spain, the remaining *Antons* served with home-based units, with most being assigned to training duties during 1939.

(Above) The twelfth production Ju 87A-1, D-IEAU, typifies the bulky but rakish profile of the 'Anton' series. The four-color camouflage of Dark Brown (61), Green (62), and Green Gray (63) splinter, over Light Blue (65) undersurfaces was the standard scheme of the period. Once the aircraft was in service, the 'civil registry' factory code was replaced by Luftwaffe unit codes and national insignia. The White rectangle on the mid-fuselage is actually light reflecting off the plexiglass first aid panel — a feature found on all Ju 87s, but which was often painted over. (US Air Force)

(Below) The floating inboard flaps are in the lowered position; the outermost sections served as ailerons. The thin metal propeller blades were highly polished metal on early 'A's, but were later finished in Black Green (70). (US Air Force)

9

(Above) One of the three Ju 87A-1s to see action in Spain with Kampfgruppe 88 of the Legion Condor, carrying the Black and White insignia of the Spanish Nationalist Air Force. As a whimsical comment on the Ju 87's appearance, this unit was dubbed the 'Jolanthe Kette' (after a popular German comedy character) and each Ju 87A-1 was decorated with 'Jolanthe the Pig' on the outside of each undercarriage trouser. The Stuka's success in Spain paved the way for being put into large scale production.

(Above) A trio of Ju 87A-2s serving with St.G 165 in 1938, displaying a not-uncommon variety of camouflage finishes — usually obtained by switching the basic colors and splinter pattern. The Red tail banner and White circle behind the swastika were soon eliminated on operational aircraft. The fuselage code on the nearest aircraft is 52+C24. (Zdenek Titz)

Tail and Rudder Development

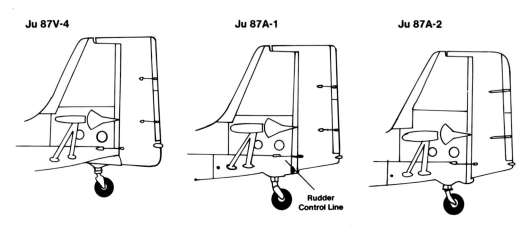

Ju 87V-4   Ju 87A-1   Ju 87A-2

Rudder
Control Line

(Right) Black-garbed Luftwaffe groundcrewmen peer from the alternate starboard hatches of an 'Anton'. The main hatches, which included the uppermost panels, swung upward from the portside along the same hinge line. The Yellow triangle beneath the pilot's cockpit indicated the use of 87 octane fuel. The White arrow below the triangle pointed to the wing root fuel filling cap (both port and starboard).

(Above) Like many training 'Antons', NG+RH has been stripped of its rear armament and has been assigned an additional number in White on its undercarriage trousers. Perhaps the greatest service performed by the underpowered Ju 87A was the preparation of young pilots for the rigors of combat, many Ju 87As survived in this capacity until well into the war.

(Above) Less than two years after its service introduction, Ju 87As were being handed over to the Stuka Schulen (Stuka Schools). Transferring the 'Antons' to schulen units resulted in some confusion in the markings found on Ju 87As with the schulen sometimes replacing the old style markings with the new style markings and sometimes retaining the older operational markings — in the foreground, a Ju 87A-2 (NG+SH) carries the new four letter code and broad-chord White on the national crosses, while the background aircraft (S13+S29) carries a non-standard variation on the old system and narrow chord White on the national crosses.

(Below) A pair of Stukaflieger (Stuka pilots) pose in front of their aircraft as the 'blackmen' prepare to crank up the inertia-starter of the Jumo 210 engine. The drooping pitot boom and radiator intake braces were unique to the 'A' series. Also unique to the 'A' series were the greenhouse mounted 'devil's horns' radio masts, antenna wires ran from their tips to the tips of the horizontal tailplanes.

Main Undercarriage Arrangement

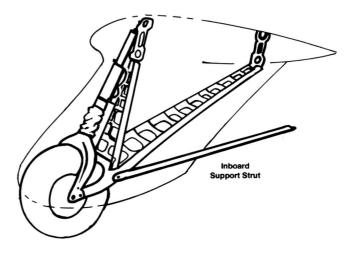

Inboard
Support Strut

11

(Above) A little known fact is that Ju 87As served in limited numbers with the Hungarian Air Force, presumably in the training role. A pair of Ju 87A-2s were also shipped to Japan in late 1937 and assembled at Mitsubishi Heavy Industries for evaluation by the Japanese Army Air Force. However, the type was never used operationally by the Japanese. (Ernest Vagi)

(Above) Color details of the Hungarian 'Antons' are unknown, but markings included the Red-White-Green tail strips, and Black and White national insignia on the wings and fuselage sides. The Hungarian Air Force would make devastating use of later Ju 87Bs and Ju 87Ds against the Soviet Army. (Ernest Vagi)

Cockpit Interior

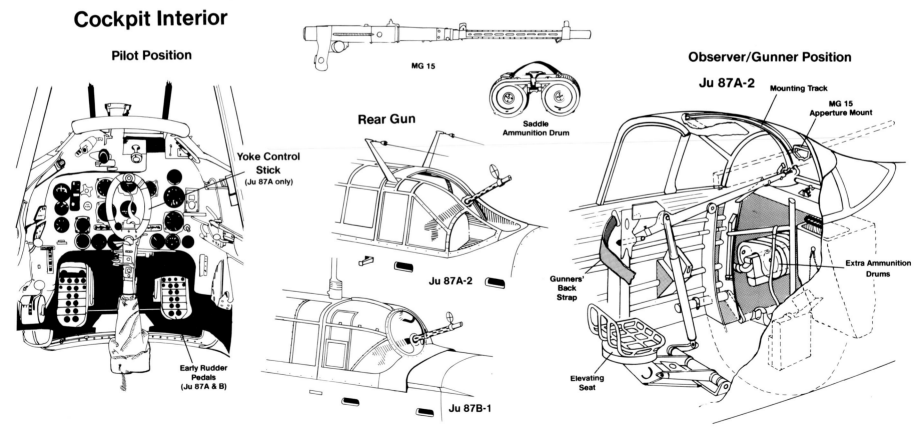

Pilot Position

Yoke Control Stick
(Ju 87A only)

Early Rudder Pedals
(Ju 87A & B)

MG 15

Rear Gun

Saddle Ammunition Drum

Ju 87A-2

Ju 87B-1

Observer/Gunner Position

Ju 87A-2

Mounting Track

MG 15 Apperture Mount

Extra Ammunition Drums

Gunners' Back Strap

Elevating Seat

Ju 87B-1

With the availability of the new 1,100 hp Junkers Jumo 211 A engine, Hermann Pohlmann's design team set out to completely overhaul the Ju 87, eliminating the shortcomings of the cumbersome *Anton*. These modifications, which were evolved on developmental prototypes Ju 87V-6 through V-9, would finally produce the first truly powerful and battle-worthy *Stuka*, under the designation Ju 87B. Major changes did not materially effect the basic airframe, but were confined to three principal areas: nose, canopy and undercarriage.

The cowling was totally redesigned to accommodate the new Jumo powerplant, including a deeper rounder radiator bath for engine cooling. The oil-coolers were consolidated into a single 'saddle' tank mounted atop the engine, with an asymmetrical air intake being added to the upper cowl with an adjustable rear outlet flap. The supercharger air intake was enlarged and relocated midway down the starboard side of the cowling.

The landing gear was completely revised, with the heavily-braced units of the *Anton* giving way to much smaller single cantilever legs without external braces. This permitted a redesign of the large ungainly exterior covers which were replaced by compact 'spats' with a telescoping upper collar to allow for oleo compression.

Crew quarters were revised featuring a new contoured four-piece greenhouse with sliding sections replacing the outward opening panel hatches. A single centerline mounted main antenna mast replaced the twin oblique angled radio masts, and a rotating rear gun mount with a pivotal ball-and-socket attachment for the gunner's MG 15 was installed. Interior improvements included a new pilot's control column, redesigned cast-magnesium overturn structure, and the replacement of the old *Stuvi* diving sight with a Revi C-12-C bombing and gun-aiming reflector sight. For the gunner, whose only duty had been that of rear defense, now came the additional responsibility of operating the main FuG VIIa cockpit radio (located behind the pilot's seat). The gunner/radio operator's position was made somewhat roomier with the removal of the seat-attached gun-elevation gear which had been necessary with the rear gun arrangement of the Ju 87A.

A simplified pitot tube replaced the large pitot boom and was moved to the starboard wing, and an underbelly mast for the winch-operated radio-telegraphy trailing antenna was added. Streamlining tips to the horizontal tailplanes were later instituted on production machines.

With the increase in power provided by the 1,100 hp Jumo 211 A engine the Ju 87B was capable of lifting twice the bombload of the Ju 87A. Maximum loaded speed was increased by nearly 40 mph, and diving speed was increased to 404 mph. The power and strength to carry heavier loads, allowed offensive armament to be increased, a second MG 17 machine gun was mounted in the port wing and two ETC 50 bomb racks were installed beneath each outer wing panel for the carriage of four 50 kg (110 pound) bombs. Loaded range, however, was reduced to approximately 385 miles due to the greater fuel consumption of the Jumo 211 A engine and increased bomb load — but the extra offensive 'punch' was considered to be worth the sacrifice in range.

Following tests with the ten pre-production Ju 87B-0s, production Ju 87B-1s began rolling out in October of 1938. Five of the initial aircraft were dispatched for combat trials with the *Legion Condor*, where they naturally overshadowed the performance of the three Ju 87As. This increase in performance together with the Third Reich's preparations for war, led to an acceleration of Ju 87 production. By March of 1939 the original RLM order for 396 Ju 87B-1s had been increased to 964, and a further Junkers' facility was tooled up at Bremen-Lemwerder to handle the increase in demand.

As deliveries progressed, production line modifications were made on the cowling of the Ju 87B-1, which, along with its primary configuration, resulted in three distinct patterns being seen in service. The initial modification consisted of movable underside radiator-bath cooling flaps being installed on a number of early production machines, (including some Ju 87B-0s). On late-production B-1s, the more efficient ejector exhaust stacks were introduced. Interestingly, these two features seldom appeared together on the same machine.

By the end of August, 1939, over 460 Ju 87s had been delivered to the *Luftwaffe*, of which over 340 formed the equipment of nine *Stukagruppen* — comprising an on-paper establishment of five *Stukageschwaders* (St.G 1, 2, 51, 76 and 77), one operational training unit (*IV (St.)/Lehrgeschwader 1*) and one naval dive-bombing staffel. The Ju 87B-1 formed the backbone of Hitler's dive-bombing force as Germany sat poised on the brink of war.

(Above) Believed to be one of the pre-production Ju 87B-0s, D-IELX carries the canopy, cowl and undercarriage changes that formed the Stuka's early wartime personality. The underwing bomb racks have yet to be installed, and the revised pitot boom is still mounted on the port wing. The pre-war four color splinter scheme is carried but would be replaced by the Black Green/Dark Green splinter scheme on production machines. (Hans Redemann)

Cowling Development

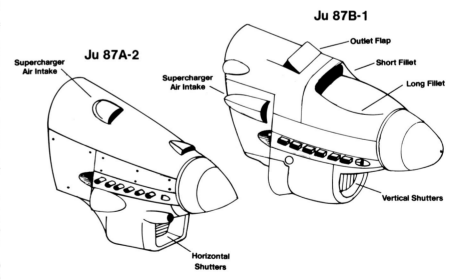

Ju 87B-1

Ju 87A-2

Supercharger Air Intake

Supercharger Air Intake

Outlet Flap

Short Fillet

Long Fillet

Vertical Shutters

Horizontal Shutters

(Above) Like its predecessor, the 'Berta' was first blooded in Spain, and with even more encouraging results. Spanish Nationalist markings were also applied to the Ju 87Bs, but the new Black Green (70), Dark Green (71) splinter, over Light Blue (65) was carried. During the six months that the 'Bertas' operated in Spain their precision dive-bombing attacks generated a virtual dive-bomber mania within the RLM. (Hans Redemann)

(Above) Ju 87B-1s sharing an assembly line with Ju 88A-1 medium bombers during 1939. In time the Junker's assembly lines at Berlin-Tempelhof and Bremen-Lemwerder would become responsible for the bulk of Ju 87 production, with only some 430 examples being manufactured at the Junkers' Dessau facility. (Bundesarchiv)

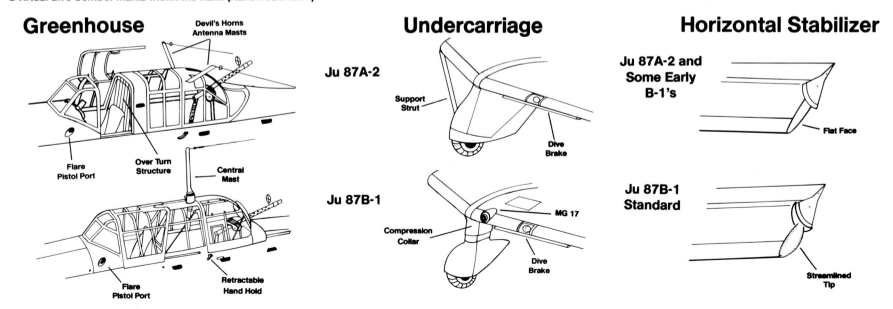

Greenhouse

Devil's Horns
Antenna Masts

Flare
Pistol Port

Over Turn
Structure

Central
Mast

Flare
Pistol Port

Retractable
Hand Hold

Undercarriage

Ju 87A-2

Support
Strut

Dive
Brake

Ju 87B-1

Compression
Collar

MG 17

Dive
Brake

Horizontal Stabilizer

**Ju 87A-2 and
Some Early
B-1's**

Flat Face

**Ju 87B-1
Standard**

Streamlined
Tip

(Above) Lehrgeschwader 1 was an operational training Geschwader of which IV Gruppe was comprised wholly of Stuka crews. The Gruppen emblem was a Light Blue shield with a Brown pitchfork-carrying devil astride a White bomb and a White letter 'L' in script painted in the forward corner of the shield. This Ju 87B-1, Coded L1+FU, has small 'F's painted in White on each wheel spat. The 'Berta's' mouth-like radiator and claw-like wheel spats added to the Stuka's mystique. The underbelly trailing antenna can be seen just behind the undercarriage, and the open vent window can be seen in the gunners' sliding canopy. (Bundesarchiv)

(Above Left) A newly delivered Ju 87B-1 has its wing guns calibrated in the angled structure of an outdoor testing range. The rotating rear gun mount vastly improved the rear gunners field of defensive fire. The early thin-chord crosses, also positioned at the extreme wingtips, would be used on Ju 87Bs until late 1939. The swastika positioned to overlap the fin and rudder would last a bit longer. (Bundesarchiv)

Diving Angle Lines

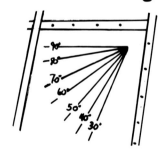

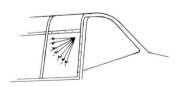

Starboard side only

(Left) Looking very much like birds of prey, 'Bertas' of St.G 51 strike the forbidding profile which would soon become the scourge of Europe — 'ugly' was a mild word for the Ju 87. The fuselage code 6G+GS is readily discernible on the nearest aircraft thanks to the glossiness of freshly-applied paint. Weather and wear will soon dull the entire aircraft to an even flatness. All letters are Black, with the third letter and spinner tip being trimmed in the 8. Staffel color of Red. The large underwing crosses would become standard for the remainder of the war. (Bundesarchiv)

The First Year Of Operations

Convinced of the timidity of the democratic nations he was facing, which had already allowed him to rearm Germany, reoccupy the Rhineland, annex Austria and absorb the Sudetenland of Czecho-slovakia, Adolf Hitler now cast a covetous eye on Poland. And as German forces began to mobilize along the Polish border, a surprise non-aggression pact was signed with Stalin on 22 August 1939, providing a free hand for the Nazi invasion of Poland without fear of Soviet intervention.

For the *Stukagruppen*, many of whose aircrews had sharpened their talents in the skies over Spain, the mission before them was essentially two-fold; precision attacks against pre-selected targets (such as radio stations, railway yards, airfields and naval bases), and direct army support strikes against enemy strongpoints encountered by advancing Wehrmacht troops and armor. The Stuka was to be the 'tip of the sword' of the *Blitzkrieg* (Lightning War), serving as 'flying artillery' whenever summoned by ground forces. In these duties it was to prove astonishingly successful.

At 04.35 hours on the morning of 1 September, three Ju 87B-1s of *3./St.G 1* dropped the first bombs of WWII when they attacked a Polish blockhouse protecting a vital bridgeway over the Vistula river — and in prophetic contrast, a *Stuka* also became the first German aircraft to fall victim to a defending fighter when a PZL P.11c shot down a Ju 87 at 05.30. However, with most of the Polish air force destroyed on the ground, *Luftwaffe* fighters quickly gained mastery of the air, allowing the Ju 87 to attack its targets nearly unopposed. During the weeks of destruction that followed, Polish forces were routed by overwhelming odds, and after the all-out bombing of Warsaw (which also involved some 240 Ju 87s) Poland capitulated on 27 September. For their part in this victory, the *Stukagruppen* lost thirty-one Ju 87s of the 285 total aircraft lost by the *Luftwaffe*. The Junkers dive bomber had proven itself to be a very economical weapon.

During the less active months that followed the fall of Poland, the so-called *Sitzkrieg* (Sitting War), the *Luftwaffe* High Command studied the effects of the Stuka, which had greatly outstripped their expectations. Average bombing accuracy fell to within thirty meters of the target, and the Ju 87 was found to be thoroughly field-reliable and popular with its crews; a prime reason for the Ju 87's popularity among aircrew was its ability to absorb heavy battle damage and return to base. As the war continued, the sight of Ju 87s hobbling home with large sections of the fuselage, tail or wings shot away, staunchly reinforced the affection which the Stuka crews held for their aircraft.

In the eyes of its enemy the Stuka emerged as a scourge, producing an unmistakable 'scream' during its diving attacks. The sound of the airstream whistling around its awkward physique and extended dive-brakes literally created panic among enemy soldiers and civilians alike. To amplify this psychological advantage, small pods were made available to be fitted on the *Stuka's* under-carriage legs, serving as mounts for optional wind-driven propeller sirens. This became an exclusive feature of the Ju 87 which would be put to extensive use.

Invasion Of The West

The uneasy calm of the *Sitzkrieg* was shattered by the spring campaigns of 1940, as German forces first pushed northward and then to the west. As in Poland, the tactics of the *Blitzkrieg* overpowered Denmark and Norway (9 April to 9 June), with *Stukas* of I/St.G 1 participating, and Holland, Belgium and France (10 May to 22 June), with Ju 87s again proving to be a decisive factor in aerial support. As British and French troops scrambled for the evacuation ships off Dunkirk in late May, *Stukas* were responsible for inflicting the greatest damage to these vessels during the few days that weather permitted *Luftwaffe* air strikes. Impressive as these victories were, however, there were also some clear warning signals. When *Stuka* formations had run into British or French fighters the Ju 87 had suffered heavy losses. As sturdy as the Stuka was, its weak rear defense, slow maneuvering speed, and lack of cockpit armor made it extremely vulnerable to fighter attacks. However since the destruction wrought by the *Stukagruppen* was out of all proportion to its losses, any serious sense of concern was ignored. Politically the *Stuka* had become a symbolic hero of the *Reich*, and its invincibility was heavily propagandized in the cinema, press and on the radio by the *Propaganda Kompanie* of Dr Goebbels. To the Allies the *Stuka* had become perhaps the most hated weapon in the Third Reich's arsenal.

(Above) 6G+AT of III/Stukageschwader 51 (later redesignated II/St.G 1) sits fully loaded and ready on a French field during the Blitzkrieg in the west during 1940. During these campaigns Ju 87 units made increasing use of the wind-driven undercarriage mounted sirens to terrorize their adversaries. This machine carries a siren-propeller mounted on the starboard undercarriage only, the port mounting is faired over. Such combinations were not unusual. (Bundesarchiv)

Undercarriage Siren

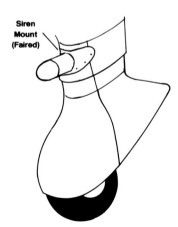

Siren Mount (Faired)

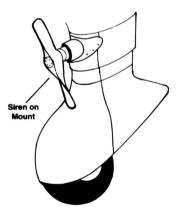

Siren on Mount

The Battle Of Britain

After the fall of France, the *Luftwaffe* quickly occupied the northern French airbases as Hitler issued both offers of peace and surrender ultimatums to Britain. With a cross-channel invasion of England (OPERATION SEA LION) seemingly imminent, the decision was made to obliterate British defenses with massive air strikes in preparation for amphibious landing.

On 6 July, the *Stukagruppen* were reorganized for the upcoming campaign. *Stukageschwader 1* was brought to full *geschwader* strength by absorbing *III/St.G 51* as its second gruppe, and the ex-naval unit *St.G 186* as its third gruppe. *I/St.G 76* became the third gruppe of *St.G 77*, bringing it to full complement, and *I/St.G 3* was prepared for action. As in Poland and France, fighter escort was to be provided by Messerschmitt BF 109s and Bf 110s. Heavy bombing duties were to be handled by the fleets of Dornier Do 17s Heinkel He 111s and Junkers Ju 88s, also escorted by fighters.

Throughout July and early August, *Stukas* successfully raided channel shipping and attacked naval facilities at Portland and Dover. But with the official declaration of *Adler Tag* (Eagle Day) on 13 August 1940, Ju 87s began their second-phase attacks against inland airfields and coastal radar stations — now pressed beyond their army-support role and flown into the thick of concentrated squadrons of RAF Hurricanes and Spitfires where *Luftwaffe* had yet to gain air-superiority. The results were disastrous. Within just six days, over forty Ju 87s were lost, while many others were forced to jettison their bombs and often returned to their bases with seriously damaged aircraft and seriously wounded crewmen. British fighter pilots had quickly learned that the very aspect of vertical bombing, which had made the *Stukas* such formidable weapons, also made them easy kills during their dives. Typical of the carnage wrought by RAF fighters was the 18 August raid against the radar installation at Poling, Sussex by *I* and *II/St.G 77*, where for only partial destruction of the station over a dozen Ju 87s were either destroyed or irreparably damaged. To their overall credit, the *Stukas* did manage to demolish several of their objectives when conditions allowed (such as Detling and Lympne airfields), but by 19 August their losses were such that they were withdrawn from operations. It was hoped that Ju 87s could be utilized in the army-support role from captured RAF bases once the planned invasion of England had been executed, but by late autumn OPERATION SEA LION was indefinitely postponed and the great *Luftangriff um England* sputtered into a humiliating failure for the *Luftwaffe*.

For the victory-starved British, the defeat of the *Stuka* — Germany's vaunted terror weapon — was a propaganda windfall. The fact that it was no more vulnerable than other dive-bombers of its time meant very little during these desperate times. *Stuka* losses were exaggerated by exuberant British politicians and stigmatized the *Stuka* as a 'Nazi hoax' and a 'sitting duck' by the Allied media. It was a reputation that would plague the Stuka for the remainder of the war and for decades afterwards. In just under a year the Junkers Ju 87 had risen to glorious heights and suffered a catastrophic defeat.

(Above) Dressed in the typical early-war bomber crew warm-weather flight suit, the gunner of S2+EN looks on as the pilot prepares to test the wing guns of his Ju 87B-1. Stukas of II/Stukageschwader 77 participated in the disastrous raids against England, and like other units it suffered tragic losses in the process. This 5. Staffel aircraft carries a Red spinner tip and Red individual aircraft letter 'E's on the front of each wheel spat. (Bundesarchiv)

(Below) A pair of underwing SC 50 (110 pound) bombs, with whistles attached to their fins also served as terror weapons. The addition of whistles was a modification made to German bombs throughout the war. (Bundesarchiv)

LUFT. FUSELAGE CODES WWII

S2	✠	A	M
▼		▼	▼
Geschwader Code		Aircraft Letter	Staffel Letter

Staff Aircraft Identification

These letters took the place of the fourth, or Staffel, symbol.

Geschwader Staff = A	III Gr. = D
I Gruppe = B	IV Gr. = E
II Gr. = C	V Gr. = F

STAFFEL IDENTIFICATION

Staffel Colour	I Gruppe	II Gruppe	III Gruppe	IV Gruppe	V Gruppe
White	1st Stfl. = H	4th Stfl. = M	7th Stfl. = R	10th Stfl. = U	13th Stfl. = X
Red	2nd Stfl. = K	5th Stfl. = N	8th Stfl. = S	11th Stfl. = V	14th Stfl. = Y
Yellow	3rd Stfl. = L	6th Stfl. = P	9th Stfl. = T	12th Stfl. = W	15th Stfl. = Z

(Above) The low contrast between Black Green (70) and Dark Green (71) led to the misconception that the uppersurfaces of Stukas were routinely painted overall Black Green, the two-tone splinter scheme of 70/71 was the standard factory finish. Staffel colors were the most frequently used additional 'in service markings'; T6+GM of 7./St.G 2 is trimmed in White, while the spinner tip of the 8. Staffel aircraft in the background is Red. Occasionally, however, Gruppe colors were employed instead. (Bundesarchiv)

Ju 87B-1 Cowling Development

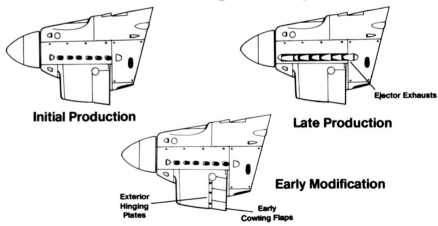

Initial Production

Ejector Exhausts

Late Production

Early Modification

Exterior Hinging Plates

Early Cowling Flaps

(Left) Thanks to some careful photo-editing, this well known photograph of L1+AU and L1+HU of 10./LG 1 makes them appear to be diving when they were actually photographed in level flight! One of the modifications made to the early production Ju 87B-1 was the temporary addition of the somewhat crude radiator cooling flaps. The aircraft in the foreground (HU) has the small hinging plates at the rear of the radiator, while the background machine is of the later production standard with a one piece 'flange' at the rear of the radiator bath and long ejector exhausts. (Ernest Vagi)

(Above) While having its engine changed this Ju 87B-1 exposes its firewall plumbing and the sturdy support box of the swinging bomb rack. The engine side of the firewall has been left unpainted natural metal, (rather than being painted RLM Gray (02), which was the usual practice with German aircraft. Total Ju 87 motor replacement could be quickly accomplished with simple hoisting gear, making the Ju 87 extremely popular with front line mechanics. (Bundesarchiv)

(Above) Stukas were occasionally called upon to perform dusk or nighttime raids — a role to which the Stuka would be assigned on a more permanent basis later in the war. This late production B-1 with ejector exhausts has been given a temporary coat of water-based Black paint on the undersurfaces, some of which has begun to wear off the undersides of the radiator housing. Sirens have not been fitted to either the port or starboard undercarriage mounts. (Bundesarchiv)

MG 17 Wing Gun Bay

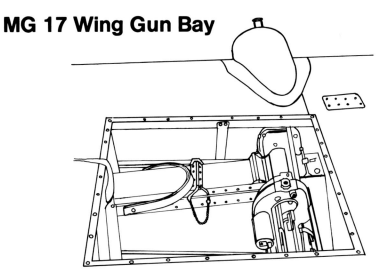

(Right) Partially concealed with foliage, a B-1 of 3./St.G 2 is prepared for crank-up during the Battle of Britain. Having seen its fair share of combat, T6+HL has become typically grubby and has had areas of repair hastily spot painted. The Junkers Jumo 211 A, not the cleanest of powerplants, has deposited heavy exhaust stains along the entire length of the fuselage. (Bundesarchiv)

Ju 87B-2

In December of 1939, the Ju 87B began receiving the uprated 1,200 hp Junkers Jumo 211 Da engine under the designation Ju 87B-2. Along with the new powerplant a number of refinements would give the B-2 its own distinct 'look', and although these represented little change over the B-1 in either configuration or performance, the Ju 87B-2 would slightly edge out the Ju 87B-1 in total numbers produced, with final revised production figures at the end of the year calling for 803 Ju 87B-1s and 827 B-2s.

The ejector exhausts and radiator bath cooling flaps, introduced as production line changes on the B-1, became standard on the B-2; the cooling flaps being redesigned for improved operation, eliminating the exterior hinging plates which were visible on early B-1s. To further increase airflow into the radiator to cool the more powerful engine, the mouth of the radiator bath was enlarged and the upper lip was raised to a shallower vee-section. For maximum propeller thrust the thin metal blades of the B-1 were replaced by broad-chord adjustable-pitch paddle blades of compressed wood. In an effort to lessen the hazard of nose-over accidents, the forward angle of the landing gear forks were increased, causing a slight enlargement in the toe sections of the exterior spats. Additional modifications included a small fresh air duct on the leading edge of the port wing, a hinged access panel over the engine starting-crank opening, and the optional use of extended covers over the muzzles of the wing-mounted MG 17s.

The Ju 87B-2 began entering operational units during the summer of 1940, too late to participate in the *Blitzkrieg* victories, but just in time to participate in the disastrous raids against England. Following this ignoble introduction, the Ju 87B-2 continued to re-equip the now less than active *Stukagruppen* as the Battle of Britain dwindled to an anti-climatic conclusion. By the beginning of 1941, the B-2 made up half the strength of the *Stukagruppen* and would soon overshadow the Ju 87B1 in service.

Umrüst-Bausätze equipment

As the Ju 87B-2 began rolling off the assembly lines in early 1940, several *Umrüst-Bausätze* (factory conversion sets) were made available in order to increase the Ju 87s versatility. When installed on the Ju 87 these *Umrüst-Bausätze* were identified by sub-designations (e.g. Ju 87B-1/U-2 or Ju 87B-2/U-2).

Umrüst-Bausätze

U-2 - Improved internal communications gear.
U-3 - Additional cockpit armor; particularly mounted to the pilot's overturn structure and the gunner's sliding canopy and rotating gun mount.
U-4 - Ski-equipped undercarriage for operations from snow-covered terrain. Although this arrangement was later service evaluated, it was seldom employed operationally.

(Right) This early Ju 87B-2 (T6+LL) of St.G 2, has yet to be outfitted with siren mounts and lacks armor plating on the pilot's overturn structure — a definite disadvantage when facing eight-gun British fighters. Stuka units continued re-equipping with the Ju 87B-2 following their slaughter over England, during which time several squadron-mates of this particular aircraft fell into British hands virtually intact. (Bundesarchiv)

Cowling Developmemt

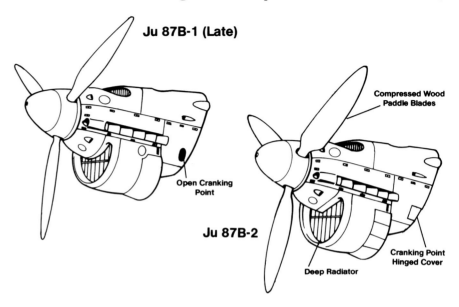

Ju 87B-1 (Late)

Compressed Wood Paddle Blades

Open Cranking Point

Ju 87B-2

Cranking Point Hinged Cover

Deep Radiator

(Above) France 1940. Armorers manhandle one of the ubiquitous three-wheeled loading carts which were used to lift bombs into position beneath aircraft. The small inscription '3. St.' on the arm of the cradle indicates that it is the property of 3./Stukageschwader 2. (Bundesarchiv)

(Below) The small duct in the leading edge of the wing between the landing light and the machine gun was unique to the Ju 87B-2 as were the hinged crank aperture cover, large compressed wood propeller blades, and the new radiator cooling flaps, seen here in the fully retracted position. This machine is S2+MM of 4./St.G 77. (Bundesarchiv)

(Above) Common to both the B-1 and B-2 was the style of the bomb swing rack, which cleared the bomb from the propeller arc after release. As the bomb is raised into position, the arms of the swing rack are adjusted to fit the load, with the tip of each arm being secured to bolts on the sides of the bomb. The suspension block atop the bomb was then locked into an opening in the Stuka's belly. Virtually without exception, the fuselage and underwing racks were finished in Black or Dark Gray. (Bundesarchiv)

Wing Leading Edge

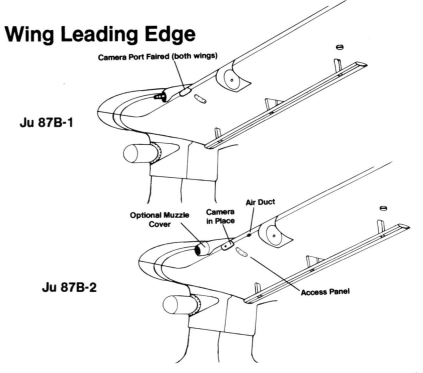

Camera Port Faired (both wings)

Ju 87B-1

Optional Muzzle Cover

Camera in Place

Air Duct

Access Panel

Ju 87B-2

(Above) The 1,200 hp Junkers Jumo 211 Da engine suspended on heavy-duty engine mounts which were secured to the firewall by four ball-and-socket mounts. The oil-cooler reservoir saddle tank can be seen above the engine, and the individual linkages for the underside cooling flaps dangle behind the radiator. The general arrangement of the powerplant was the same for both the B-1 and B-2. (Bundesarchiv)

(Left) Personnel of St.G 77 prior to a mission. This Ju 87B-2 has been fitted with complete cockpit armor, with the pilot's position being virtually walled off from the rest of the compartment, and vision from the gunner's position being reduced to a rectangle of armored glass in the center of the circular gun mount. Angled armor sheet plating lines the lower rear corners of the gunner's windows, and was often painted RLM Gray (02). This additional cockpit armor on both Ju 87B-1s and B-2s was introduced with the U-3 factory conversion kit and quickly became standard. (Bundesarchiv)

(Below) This damaged Norwegian based Ju 87B-2 is loaded onto a wooden sled to be towed back to base for repairs. Perhaps with an eye toward trying to prevent this type of mishap, the forward coverings had been removed from the undercarriages, however, soft earth, sand or mud could still cause the nose heavy Stuka to flip over. The compressed wood VS-11 propeller blades have sheared off. (Bundesarchiv)

Armored Overturn Structures

Early Overturn Structure without Armor

Later Overturn Structure with Armor

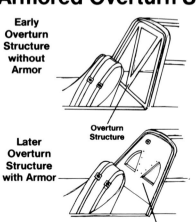

Overturn Structure

Overturn Structure

Full Armored Overturn Structures

Plate Closed

Plate Open

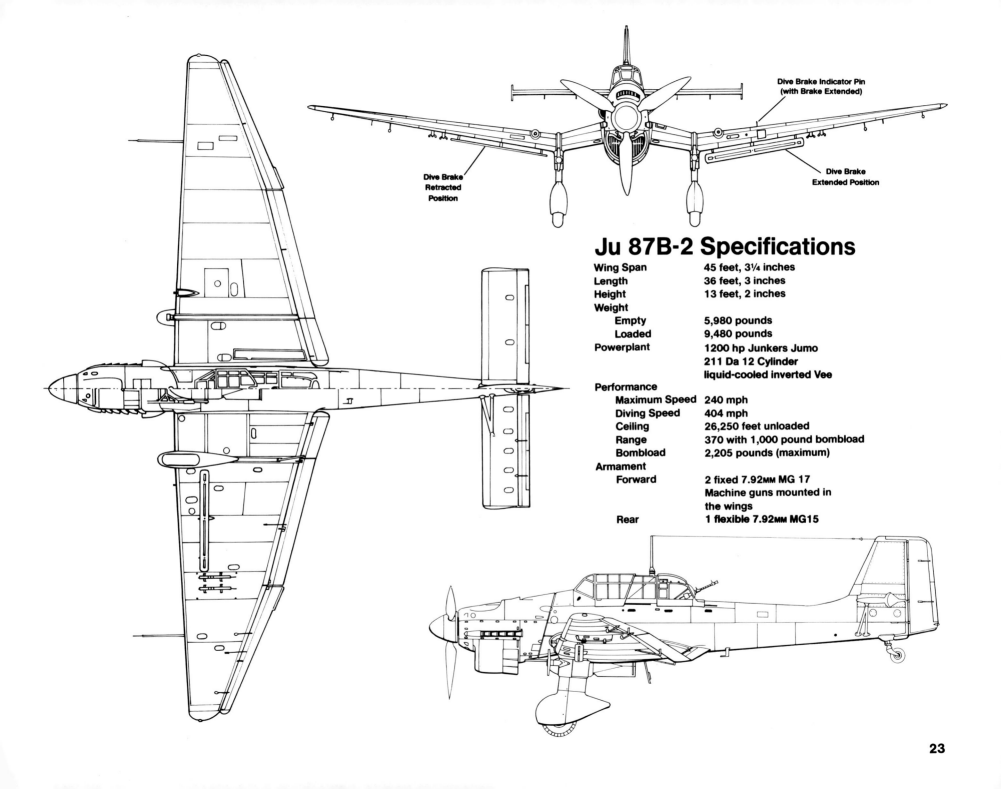

Dive Brake Indicator Pin
(with Brake Extended)

Dive Brake
Retracted
Position

Dive Brake
Extended Position

Ju 87B-2 Specifications

Wing Span	45 feet, 3¼ inches
Length	36 feet, 3 inches
Height	13 feet, 2 inches
Weight	
Empty	5,980 pounds
Loaded	9,480 pounds
Powerplant	1200 hp Junkers Jumo 211 Da 12 Cylinder liquid-cooled inverted Vee
Performance	
Maximum Speed	240 mph
Diving Speed	404 mph
Ceiling	26,250 feet unloaded
Range	370 with 1,000 pound bombload
Bombload	2,205 pounds (maximum)
Armament	
Forward	2 fixed 7.92ᴍᴍ MG 17 Machine guns mounted in the wings
Rear	1 flexible 7.92ᴍᴍ MG15

Ju 87R

As a simple solution to the poor range of the Ju 87B, the Ju 87R featured a revised internal fuel-transfer system and the ability to carry two 'strap-on' 300 liter (66 Imperial gallon) fuel tanks replacing the outer wing bomb loads, with four long pegs bracing the tanks. Despite its separate alphabetical designation, the R series (standing for *Reichweite* or range) was actually a sub-variant of the Ju 87B that simply had the ability to carry wing mounted drop tanks. When the tanks were not required for long-distance missions, the Ju 87R could be rigged to carry the standard underwing bombload, which made it indistinguishable from the Ju 87B. When bombs were carried the four braces were deleted.

Ju 87R-1

The first production variant, the Ju 87R-1, was a sub-variant of the Ju 87B-1 with ejector exhausts, owing to its later introduction on the assembly lines in 1939. The first operational unit to receive the Ju 87R-1 was *I/Stukageschwader 1* during operations in Norway, where, with its over 870 mile range the R-1 enjoyed an exceptional radius for combat patrols and enjoyed a tremendous success as a land-based anti-shipping interceptor. Because of this, the Ju 87R-1 was initially shrouded in secrecy, and early photographs of these aircraft often had the underwing tanks scratched out by German censors. Following its Norwegian introduction, the Ju 87R-1 was supplied to St.G 2 and was used in the French campaign, and the Battle of Britain. Afterwards, other Geschwadern received long-range "R s."

Ju 87R-2

The later Ju 87R-2, was based on the Ju 87B-2 airframe. External details between the B-2 and R-2 were identical, but maximum flying weight could be stressed to between 10,362 pounds and 12,456 pounds. Maximum range of the heavier Ju 87R-2 was rated below that of the R-1 at 779 Miles (180 mph outbound and 205 mph inbound, at 13,124 feet), but did not seriously inhibit the Ju 87R-2's operational career, which made its debut in the Mediterranean Theater in early 1941.

Ju 87R-3

The Ju 87R-3 was a limited production variation of the R-2, featuring improved radio gear. However, at least one source has suggested that the Ju 87R-3 was specifically delegated to the long-range glider-towing role. (See page 35)

300 Litre Fuel Tank

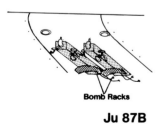

Bomb Racks

Ju 87B

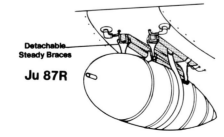

Detachable Steady Braces

Ju 87R

(Above) The waistline securing strap of a 300 liter (66 Imperial gallon) tank is shackled into place below the wing rack of a Ju 87R. Two long detachable steadying pins on each wing rack nestled into corresponding sockets on the tank uppersurface. Stenciling below the wing leading edge reads "Spannband nach dem Tanken nachziehen" (Tighten strap after filling tank). (Bundesarchiv)

(Above) Without its underwing tanks the Ju 87R was essentially a Ju 87B, and with its tanks the Ju 87R was essentially a long-range Ju 87B. When not carrying its tanks, a full bombload could be carried by the Ju 87R with only minor revision to the ETC 50 wing racks. This red-spinnered Ju 87R-1 (A5+HK) of St.G 1 carries the White, Yellow and Black 'diving crow' emblem of I gruppe on the upper cowling. (Bundesarchiv)

(Below) While based in Norway I/St.G 1 was the first unit to receive the Ju 87R-1. Due to the drag-producing underwing tanks, 'R's were usually somewhat cleaner aircraft and were seldom rigged with undercarriage sirens. Addition of the under wing tanks meant reducing the offensive load to the centerline bomb. (Bundesarchiv)

(Below) A pair of Ju 87R-2s of 2./St.G 3 returning from a patrol out into the Mediterranean in early 1941. The Med was a region in which the 'R' was repeatedly used against British convoys. Note the physical likeness between the R-2 and B-2. (Bundesarchiv)

1941

As 1940 drew to a close, the potential disaster of an Allied victory over embattled Italian forces in the Mediterranean and North Africa, plus Nazi plans to invade communist Russia to the East guaranteed that the replenished *Stukagruppen* would again be committed to battle. Despite the Ju 87's embarrassing performance during the Battle of Britain, the *Luftwaffe* still regarded the Ju 87 as an effective weapon, so long as it was used under conditions of German air-superiority and in the close tactical-support role.

Following the initial January and February bombing campaigns against Britain's Mediterranean island fortress of Malta (where Ju 87s inflicted heavy damage on the aircraft carrier HMS ILLUS-TRIOUS), *Stukas* later participated in the sweeping victories over Yugoslavia and Greece (6 to 27 April), the airborne invasion of Crete (20 May to 1 June), and the support of General Irwin Rommel's *Afrika Korps* across the deserts of North Africa. Along the vital Mediterranean waterways, Ju 87s (particularly the long-range R series) literally became the bane of English shipping. Finally, on 22 June, Ju 87s spearheaded the thrust into the vast expanse of the Soviet Union. Before long, however, it became obvious that the 'rapid subjugation' of the East was turning into a meat-grinder, which would sap Germany of its men, material and strength.

After 1941, the fortunes of war and the excesses to which Hitler had pushed his forces would begin to erode the power of the Third Reich — but not before the Junkers Ju 87 had re-established itself as the war's most efficient dive-bomber.

(Below) With their belly racks empty Yellow nosed Ju 87B-1s of II/St.G 77 return from sortie. OPERATION MARITA, the campaign through Yugoslavia and Greece, was a stunning success for the dive bomber and did much to resurrect the Stuka's reputation, which had been so tarnished during the Battle of Britain. Casualty figures in the German wartime publication 'Wir Kampften auf dem Balkan' lists Stukageschwader crew losses from 6 April to 1 June 1941, as 42 killed or missing — far less than the losses suffered by the Kampfgeschwadern (bomber units). (Bundesarchiv)

(Above) The undercarriage siren, cowl, and stencil markings can be appreciated here, including the small manufacturer's data plate just forward of the wing root. The Yellow, Red and Black shield of Stab/II/St.G 77, which saw action in the Balkans and Mediterranean during the early months of 1941, is carried below the windscreen. (Bundesarchiv)

FuG 25 and Peil G.IV

In 1940 the FuG 25 I F F (Identification Friend or Foe) radio and the Peil G.IV direction-finder became available for fitting to the Ju 87B-2 and R-2, as well as retrofitting to some B-1s and R-1s. The FuG 25 was detectable by a small centerline whip aerial beneath the aft fuselage, and remained an optional feature throughout all subsequent Ju 87 variants. The unique Peil G.IV featured a belly-mounted cylindrical loop-antenna beneath the gunner's station, which was covered by a small semi-transparent protective plexiglass dome. A late 1941 British Air Ministry report on a captured Italian Ju 87B-2 recorded a further interesting detail concerning the dome itself:

> *the interior of this cover is painted in strips with metalizing paint. A cross strip joins these together electrically... This is evidently some capacity earth effect or more probably the sense aerial of which it is a particularly ingenious form..*

Photographs indicate that both of the above installations were installed on some aircraft from early 1941 onwards, but certainly not all Ju 87B-2/R-2 aircraft were so equipped. Apparently, whenever the Peil G.IV was installed, the belly telegraphy trailing-antenna mast was removed.

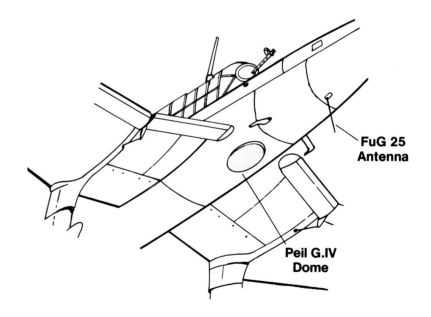

FuG 25 I.F.F. and Peil G.IV D/F

FuG 25 Antenna

Peil G.IV Dome

(Above) A Ju 87R-2 of 2./St.G 3 presides over the wreckage of RAF aircraft on a conquered Greek airfield. April 1941. (Bundesarchiv)

(Above) Occasionally Ju 87s were assigned to units other than the Stukagruppen. A close examination of the nearest Ju 87B-2 reveals the code 2F+CA, indicating that it is on charge with either KG 54 or ZG 26. Pomp and ceremony, and the display of Luftwaffe regalia was a common sight on German airbases, as was the photo-snapping Propaganda Kompanie (PK) officer. (Bundesarchiv)

Ju 87B/R Tropical

With the introduction of the *Stuka* into the dusty, arid climate of the Mediterranean area including North Africa, installation of desert survival equipment and supercharger dust filters became a necessity. Ju 87s receiving the desert modification were given the additional designation *trophische* (tropical), usually abbreviated to *trop*. This resulted in four additional Stuka sub-types:

**Ju 87B-1/trop and Ju 87B-2/trop
Ju 87R-1/trop and Ju 87R-2/trop**

Although the various *Stukagruppen* assigned to the Mediterranean area were initially equipped with all four types, the Ju 87B-2/trop and Ju 87R-2/Trop quickly became the most numerous variant.

Ju 87R-4

The principle desert variant was the Ju 87R-4, which was identical to the Ju 87R-2/trop, but incorporated the climatic modifications on the assembly line. When not equipped with the 300 liter underwing tanks, the R-4, like the R-2/trop, was indistinguishable from the Ju 87B-2/trop.

(Right) During the opening phases of North African involvement, Tropicalized Ju 87Bs and 'Rs' were flown into battle with the splinter Green camouflage and early broad-chord White theater bands on the rear fuselage. T6+FM of 4./St.G 2 carries the emblem of the Deutsches Afrika Korps (DAK) on its starboard cowling; a relatively short-lived marking indicating that they were part of General Rommel's force. In addition to the White fuselage band theater markings would include White undersides to the wingtips. (Bundesarchiv)

Air Filter

Standard Air Filter	Desert Dust Filter

 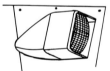

(Middle Right) Later, as the desert war became more intense, Sand-Yellow desert-terrain camouflage was introduced in a variety of patterns to the upper surfaces, in this case temporarily covering the unit code on the fuselage. Early Sand-Yellow was often field-mixed, conforming to no official color specification, and was applied both with brush and spray-gun. (Bundesarchiv)

(Right) Production aircraft bound for the desert carried the factory applied official camouflage upper surfaces of Sand-Yellow (79) over Light Blue (78) undersurfaces; contrast between the two colors was such that little difference was seen in black and white photos. Blotches of Olive Green (80) were usually sprayed on the uppersurfaces at unit level. Factory codes indicate that this Ju 87R-4 was probably at a Luftwaffe airpark prior to delivery to an operational unit. (Bundesarchiv)

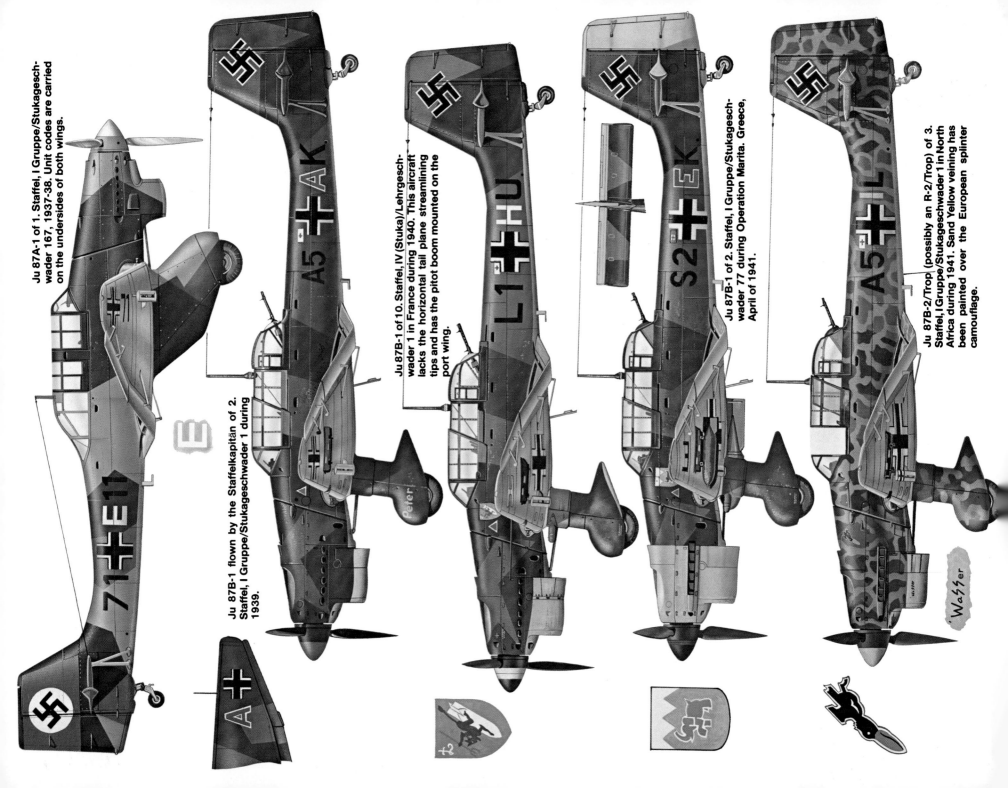

Ju 87A-1 of 1. Staffel, I Gruppe/Stukageschwader 167, 1937-38. Unit codes are carried on the undersides of both wings.

Ju 87B-1 flown by the Staffelkapitän of 2. Staffel, I Gruppe/Stukageschwader 1 during 1939.

Ju 87B-1 of 10. Staffel, IV (Stuka)/Lehrgeschwader 1 in France during 1940. This aircraft lacks the horizontal tail plane streamlining tips and has the pitot boom mounted on the port wing.

Ju 87B-1 of 2. Staffel, I Gruppe/Stukageschwader 77 during Operation Marita. Greece, April of 1941.

Ju 87B-2/Trop (possibly an R-2/Trop) of 3. Staffel, I Gruppe/Stukageschwader 1 in North Africa during 1941. Sand Yellow veining has been painted over the European splinter camouflage.

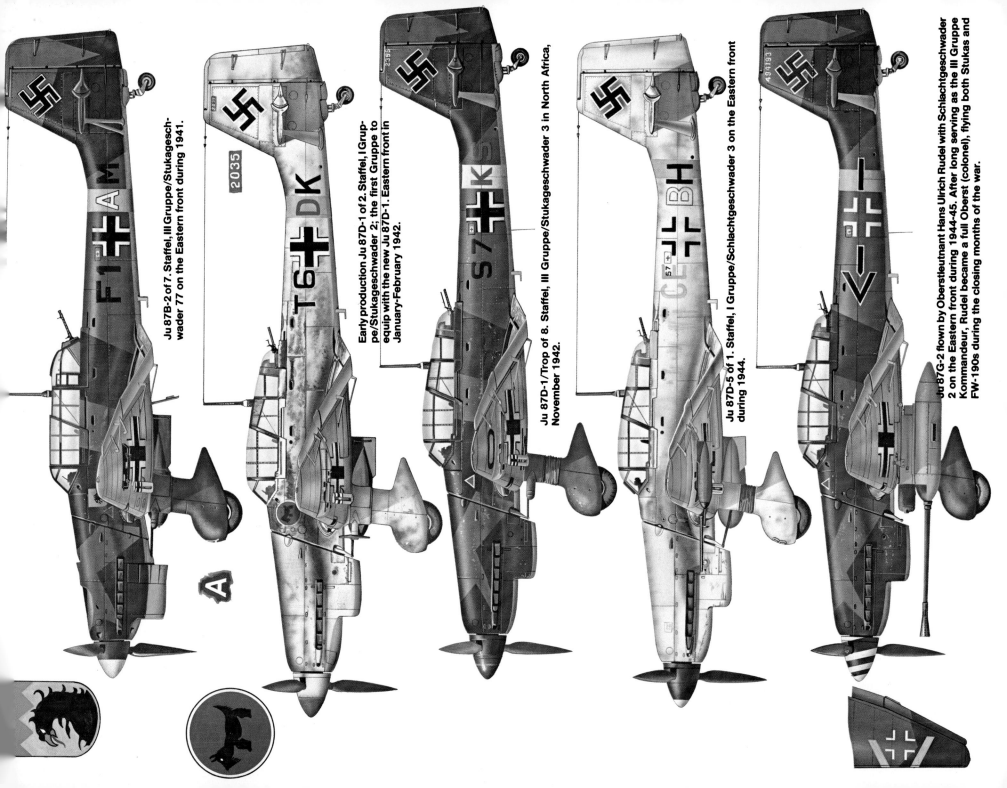

Ju 87B-2 of 7. Staffel, III Gruppe/Stukageschwader 77 on the Eastern front during 1941.

Early production Ju 87D-1 of 2. Staffel, I Gruppe/Stukageschwader 2; the first Gruppe to equip with the new Ju 87D-1. Eastern front in January-February 1942.

Ju 87D-1/Trop of 8. Staffel, III Gruppe/Stukageschwader 3 in North Africa, November 1942.

Ju 87D-5 of 1. Staffel, I Gruppe/Schlachtgeschwader 3 on the Eastern front during 1944.

Ju 87G-2 flown by Oberstleutnant Hans Ulrich Rudel with Schlachtgeschwader 2 on the Eastern front during 1944-45. After long serving as the III Gruppe Kommandeur, Rudel became a full Oberst (colonel), flying both Stukas and FW-190s during the closing months of the war.

(Above) Conditions in North Africa were often bleak and primitive. This Ju 87R-2/Trop (under-wing tanks can be seen on the ground next to the revetment of drums and sand bags) wears a foul-weather tarp. These tarps were standard issue with each Ju 87, and were often labeled with corresponding aircraft serial numbers. (Bundesarchiv)

(Above) In the absence of the three wheel hydraulic loading cart, bombloads could be hoisted into place by using block and tackle gear and old fashioned 'elbow grease'. The hinged panel being opened by the ground crewman provided access to the upper fuselage oil tank for servicing. The smaller of the two holes in the port wing leading edge indicates that this Ju 87B-2/Trop has been fitted with an auto-camera. (Bundesarchiv)

(Below) While taxiing out for a mission, S1+GK of 2./St.G 3 suffered a Kopfstand (headstand), which was one of the less fortunate traits of the somewhat nose-heavy Ju 87. When landing, Stuka pilots were advised to make a three-point touch down whenever possible. Overall Sand-Yellow uppersurfaces were not unusual on early desert aircraft of St.G 3. (Bundes-archiv)

(Below) 2F+CA, a Ju 87 R-2/Trop with Sand Yellow sprayed over the splinter Green upper surface. With desert 'R's flying increasingly short-range missions against British forces, the 300 liter tanks became less frequently used. Beneath the aft fuselage is the antenna for the FuG 25 IFF radio, and the plexiglass bubble of the Peil G.IV D/F aerial is just visible between the inboard flaps. (Smithsonian Institution)

(Above) Italy received over 100 Ju 87s and used them effectively in the Mediterranean and North Africa. This Regia Aeronautica Ju 87B-2/Trop (Werke N. 5763) fell into British hands in September of 1941 and was the subject of an Air Ministry report concerning its Peil G.IV direction-finding device. Uppersurfaces have been resprayed overall Dark Green, with Yellow nose and White Italian national insignia. (Imperial War Museum)

Ju 87B/R Rear Gun Development

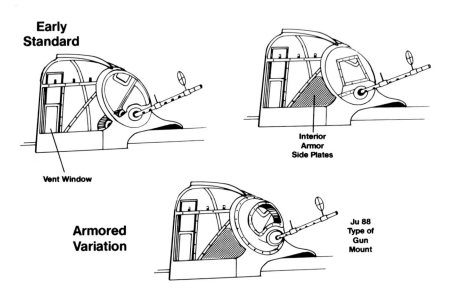

Early Standard

Vent Window

Interior Armor Side Plates

Armored Variation

Ju 88 Type of Gun Mount

(Above) Luftwaffe officers chat beside a Ju 87B-2/Trop of St.G 3 after a meeting with Field Marshall Albert Kesselring at El Adem on 19 June 1942. The gunner's position is equipped with an armored gun mount of the type more commonly associated with the upper rear defensive positions of the Ju 88. This same rear gun installation was found on Ju 87Bs and Rs serving on other fronts, and may have been a late factory fitting. The gun is complete with a sun shade over the gunsight. (Bundesarchiv)

(Below) By the spring of 1942 St.G 3 had absorbed I/St.G 1 as its second gruppe and II/St.G 2 as its third gruppe. The revised 'S7' Geschwader code of St.G 3 replaced the 'A5' Geschwader code on the fuselage of St.G 1 aircraft, but the gruppe continued using its diving crow emblem throughout the remainder of the desert war. The pilot's belly viewing-window structure has been removed and can be seen in the the foreground with the upper adjustment handle projecting to the right. (Bundesarchiv)

RUSSIA

(Above) With the opening of hostilities against Russia on 22 June, Stukas were unleashed with destructive precision. This mixed flight of Ju 87B-1s and B-2s of II/St.G 1 are returning after hitting targets. (Bundesarchiv)

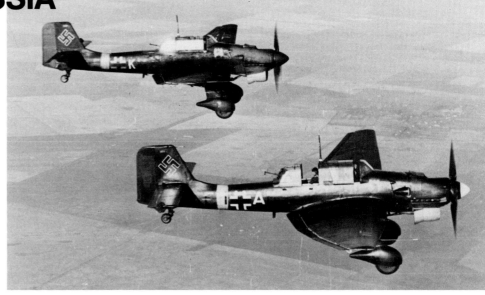

(Above) Although sources have indicated that I/St.G 76 was reassigned as III/St.G 77 in 1940, aircraft within the former I/St.G 76 retained their original geschwader code of 'F1' and a second-gruppe coding structure well into the war. F1+KM of the 7. Staffel carries the Yellow fuselage band and udersides of the wingtips which were the official theater markings of the Eastern Front. (Bundesarchiv)

(Below) The Stuka at work! A Ju 87 pulling out of its dive has just struck a direct hit on Soviet bridgework. However, the slow tedious pullout from its dive was the point at which the Stuka was most vulnerable — something Allied fighter pilots quickly learned to exploit. (Bundesarchiv)

(Below) Open fields or roadways were all the same to the Stuka as it leapfrogged behind the ever-moving Wehrmacht. The stacked SC 50 bombs have been staggered in their piles to prevent damage to the percussion detonators while they await loading. The extended fuses were used to detonate the bombs above ground, scattering shrapnel amongst Soviet troops. (Bundesarchiv)

(Above) During the war in the East, elements of IV (Stuka)/Lehrgeschwader 1 would be operated extensively in Finland, and later helped to form St.G 5. It is believed that some Stukas received by this unit were factory equipped desert aircraft, diverted to the Russian Front and repainted with Dark Green (71) uppersurfaces, leaving traces of Sand Yellow (79) behind stenciling points and other difficult to paint areas of the aircraft. (ECP Armee)

(Above) L1+DW takes on fuel in its starboard wing root tank, the rubber bumpers on the hose helps to prevent paint from being scraped off the wings. Again, small Sand Yellow patches are visible, as well as a tropical supercharger air intake dust filter on the cowling. Undersurfaces are probably desert Light Blue (78). (ECP Armee)

(Below) As Hitler's Lightning War in the East dragged on into the winter of 1941, Luftwaffe aircraft were snow-camouflaged with temporary water-based White distemper. With a bit of artistic flare, this Ju 87B-2 sports a sharkmouth on the cowling and tiger stripes on the undercarriage spats! (Bundesarchiv)

(Below) The Ju 87B-2/U-4 conversion offered a logical solution to operations from snow and ice-covered fields, but it did not see widespread use. Nevertheless, ski equipment remained a planned option on subsequent Ju 87 variants. (Bundesarchiv)

Ju 87B/R Glider Tug

(Above) Along with its usual bombing duties, the sturdy Ju 87 was found to be an ideal light glider towing aircraft. Simplified markings on these Ju 87B-1 glider tugs may indicate that they are serving as home-based trainers. The Werke number on the aircraft in the foreground is 5222, and carries a Yellow fuselage band and rudder number. The spinner tip appears to be Red edged with White. (Bundesarchiv)

(Above) The principle load of Ju 87 glider tugs was the DFS 230A or B cargo/troop glider, here being haulded aloft by a Ju 87R-2. The Ju 87R series were particularly handy in the tug role because of their range, and were used extensively in the Mediterranean and Russia. (Bundesarchiv)

Glider Towing Installation

(Below) A groundcrewman attaches a towing cable to the strong, simple tail-box attachment point which was common to all Ju 87 tugs. The external tubular mounting frame was fastened to the aft fuselage bulkheads, and could be field-installed. (Bundesarchiv)

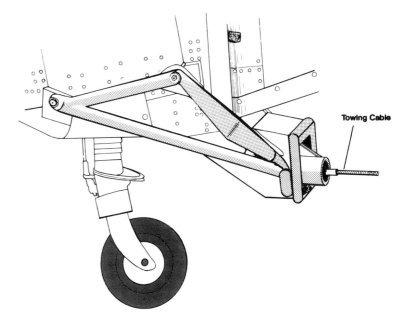

Towing Cable

Ju 87C Carrier Dive Bomber

With the preparation of a dive-bombing gruppe for use aboard the future *Kriegsmarine* aircraft-carrier *Graf Zeppelin*, 4./*Trägergruppe 186* (Carrier Group 186) was activated in December of 1938 with Ju 87As as initial equipment. However, with the introduction of the improved Ju 87B, plans for creating a fully navalized *Stuka* were made under the designation Ju 87C. The Ju 87C was a B-variant modified with catapult gear, an arrester hook beneath the aft fuselage, jettisonable undercarriage for emergency ditching, and manually operated rearward folding wings for storage below deck. Wing span was cropped to 43 feet 3 inches for easier deck handling. When folded, the wing span was reduced to roughly that of the horizontal stabilizer at 16 feet 5 inches.

During March and April of 1939, two early Ju 87Bs were converted as Ju 87C prototypes, with ten pre-production Ju 87C-Os being produced at Tempelhof during the summer. These initial navalized *Stukas* are believed to have been issued to 4./*Trägergruppe 186* for service evaluation, and may have seen action in Poland alongside the Ju 87B-1s known to have been operated by the unit. Unfortunately the Ju 87C would never be used in its intended role, that of a seafaring dive-bomber, since construction of the *Graf Zeppelin* had been brought to a standstill by the end of the year; postponed by what were considered to be more practical wartime projects.

Only a few production Ju 87C-1s (based on the Ju 87B-2) had been completed before recognized obsolescence of the B-airframe brought the cancellation of the order for 170 navalized *Stukas* in 1940 and spelled the end of the Ju 87C series. Nevertheless, experiments with these aircraft, including flotation, catapult and armaments trials, continued until 1942 with the possibility that the *Graf Zeppelin* project might be completed at a later time.

4./*Trägergruppe 186* was brought to full gruppe strength and integrated into the Luftwaffe — being referred to as *Stukageschwader 186*. In July of 1940 it became *III/Stukageschwader 1*. Its personnel continued to decorate their Ju 87s with their original anchor-and-helmet unit shield — a colorful reminder of what might have been.

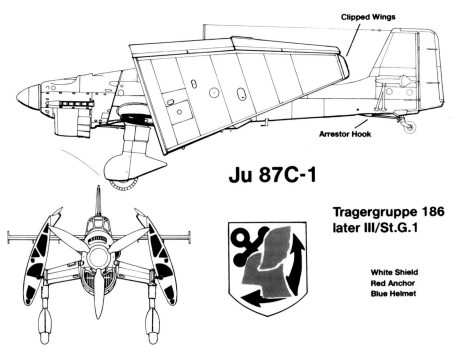

Ju 87C-1

Tragergruppe 186 later III/St.G.1

White Shield
Red Anchor
Blue Helmet

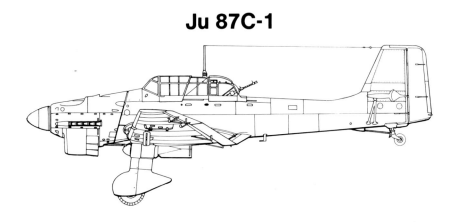

Ju 87C-1

(Right) Often described as being an early Ju 87B-1, it is believed to be one of the Ju 87C-O aircraft carrier prototypes. The early type of arrester hook arrangement is visible beneath the aft fuselage, as is the research boom on the starboard wing. The aircraft is natural metal with the Red, White and Black Swastika tail banner. The few production aircraft produced were camouflaged in the standard 70, 71, 65 scheme. (Zdenek Titz)

Ju 87D Development

During the aftermath of the Polish campaign, the weak defensive protection of the Ju 87 and its inability to breech extremely well fortified targets became apparent to the *RLM Technische Amt*. While a future generation of advanced attack aircraft was envisioned (such as the twin engine Messerschmitt Me 210 and the Jumo 213 powered Ju 187), an improved interim Ju 87 would be needed pending their development. During 1940 a total renovation of the Stuka was ordered under the designation Ju 87D. Project goals were aerodynamic refinement, the ability to carry a heavier, more penetrating bombload, increased aircrew protection, and stronger defensive fire. The Ju 87D prototype was originally planned for flight-testing in December of 1940. However, due to the troublesome Junkers Jumo 211 F engine, the first completed prototypes (Ju 87V-21, V-22 and V-23) could not be flown until the more reliable 1,400 hp Junkers Jumo 211 J became available in February of 1941. Again the major areas of change were focused on the nose, canopy and undercarriage.

The entire cowling was aerodynamically re-contoured and slightly lengthened, the oil-cooler was relocated to a shallow bath on the underside of the cowling; the coolant system was divided and relocated into two radiators, one under each inboard wing. The starboard mounted supercharger air intake was moved farther forward and streamlined into the cowl surface.

To further reduce drag, the canopy was streamlined into a more rounded, rearward-sloping profile, with the gunner's sliding section incorporating a GSL-K 81 armored turret which manually rotated under the open end of the canopy. Defensive armament was upgraded with a 7.92MM MG 81Z *zwilling* (twin) machine gun mount with 2,000 rounds of belt-fed ammunition from two cockpit floor lockers. Interior cockpit armor was increased along the sides, pilot's seat, floorboard and rearmost cockpit bulkhead.

The undercarriage was strengthened to handle increased weight, while the coverings were once again reduced in outline with smaller spats, thinner leggings, and leather-covered oleo compression bands. The Peil G.IV D/F antenna became a standard installation, being relocated in a recessed well in the upper rear fuselage deck and sealed beneath a flush plexiglass cover. The vertical tail was slightly increased, and the horizontal tail "vee-braces" were replaced by aerofoil-section braces.

For greater variation of payload, the fuselage bomb rack was completely redesigned, and new streamline-covered all-purpose racks were installed beneath the wings. Each of these three-in-one wing racks were capable of carrying a pair of side-by-side small caliber bombs, or a single larger bomb on a center slip. Ordnance options now included AB 250 or AB 500 wooden dispensers (for incendiaries or SD-2 and SD-4 anti-personnel bombs), machine gun packages (*Waffenbehälter WB 81A or B*) for greater strafing power, all-purpose containers (500 liter capacity), or flare and smoke discharger tubes.

Increased performance of the Ju 87D included a maximum unloaded speed of 255 mph, an optimum short-range bombload of 3,968 pounds, and a maximum range of over 500 miles with additional internal wing fuel tanks. For increased versatility the Ju 87D could be outfitted with snow skis, underwing 300 liter fuel tanks (which extended range to 953 miles), and could be readily adapted for desert use under the *trop* sub-designation.

As the Ju 87D made its appearance on assembly lines during the summer of 1941, overall production of the Ju 87 was in the process of winding down. The new *Dora* may have been a vast improvement over the Ju 87B, but it also represented the limit to which the airframe could be developed. But as it became obvious that the long awaited 'new breed' of dive-bombers would not soon materialize, production of Ju 87 was greatly accelerated to meet the needs of the *Stukagruppen*. As a result some 3,000 Ju 87Ds (and direct adaptations) were built between 1941 and 1944. With total *Stuka* production amounting to an estimated 5,709 aircraft, the Ju 87D series was therefore the most numerous. Unfortunately the *Dora* was used to fight a largely defensive war along Germany's declining frontlines, and would never capture the highly publicized glories of the *Berta*.

Ju 87D-1

During early flight trials weaknesses in the new landing gear forks and mainwheels of the Ju 87D caused a number accidents due to undercarriage failures. To solve this problem initial production Ju 87D-1s were fitted with the proven landing gear and coverings of the Ju 87B-2 (restricting maximum

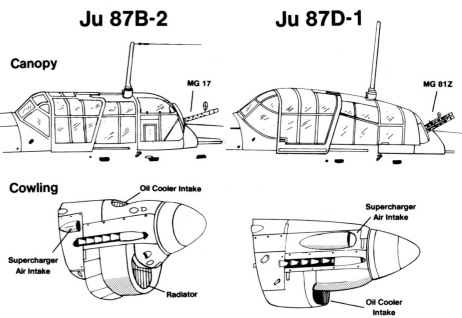

Ju 87B-2   **Ju 87D-1**

Canopy — MG 17 — MG 81Z

Cowling — Oil Cooler Intake — Supercharger Air Intake — Supercharger Air Intake — Radiator — Oil Cooler Intake

Ju 87D-1 Underwing Radiators

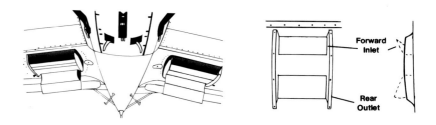

Forward Inlet — Rear Outlet

loaded weight from 14,550 pounds to 12,787 pounds). Late production D-1s introduced a strengthened version of the new undercarriage design, but landing gear problems would continue to plague the Ju 87D throughout its career.

The Ju 87D-1 first entered service with I/St.G 2 on the Russian front in January of 1942, replacing the Ju 87B and R variants as supplies of the new machine became available. While the greater striking power of the *Dora* was certainly welcomed, the cramped quarters of the Ju 87D cockpit was disliked by its crews. And while certain seating adjustments were made for the pilot, the gunner's low-profile station remained unchanged, and with its new arrangement of armor plate, ammo boxes and larger breeches of the MG 81Z, it was decidedly uncomfortable in comparison to the Ju 87B.

Progressive combat reports led to further production line changes, mainly aimed at increasing crew safety. A 50MM armored glass windshield quickly became a standard fitting, and the Ju 87D-1 became the first *Stuka* to feature optional exterior armor plate around the central fuselage, and both inside and outside of the canopy. These applications of armor plate were at first quite extensive, but later were limited to only the most practical areas of needed protection (the fuselage sides beneath the pilot's cockpit and the lower forward fuselage behind the cowling).

A total of 592 Ju 87D-1s were completed during 1942, and most would be swallowed up in the mounting holocaust along the Russian and North African fronts by early the following year.

(Right) The first unit to receive the Ju 87D-1 was I/St.G 2 in January of 1942; in this case a whitewashed aircraft of 1. Staffel. The retro-fitting of the stocky-looking Ju 87B-2 type of undercarriage was a relatively short-lived interim feature, characterizing the first combat 'Doras'. (Bundesarchiv)

Undercarriage Development

Ju 87D-1
Early

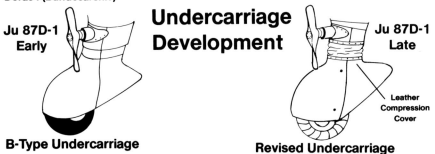

B-Type Undercarriage

Ju 87D-1
Late

Leather
Compression
Cover

Revised Undercarriage

(Below) St.G 1 was the second unit to receive the Ju 87D-1, following closely behind St.G 2. Shortly after their introduction to combat, the early D-1s were joined by machines with the revised undercarriage design which can be seen on the aircraft in the background. Heavy exhaust stains have darkened the panels of the applique armor on the forward fuselage of the nearest machine. (Bundesarchiv)

(Left) Early production Ju 87D-1s taking off from a snow-covered airfield — or as the Germans sometimes referred to it, 'flying from a tablecloth'. February of 1942. (Bundesarchiv)

Appliqué Armor

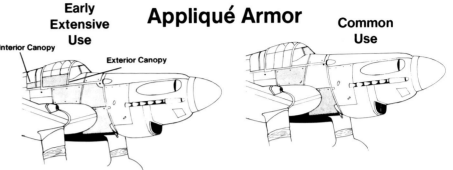

Early
Extensive
Use

Interior Canopy

Exterior Canopy

Common
Use

(Below) This atmospheric pre-dawn scene may look attractive, but the harsh Russian winters were incredibly brutal to both men and machines. Day after day sub-zero temperatures froze oil and fuel lines and cracked propeller blades — problems which were only partially solved with heating tents over the Stukas' cowling. Another hazard was the deep hatred that Soviet troops had for both the Stuka and its air crews, making the prospect of bailout behind enemy lines almost unthinkable. (Bundesarchiv)

(Above) The greenhouse of the 'D' series was more rounded in both profile and cross section than the 'B' series. The 'Dora's' redesigned canopy also afforded the pilot substantially more armor protection with his seat-mounted headrest and the overturn plating (just visible at the front of the center canopy section). On all variants of the Stuka from the Ju 87B onward nearly half the canopy framework was on the inside of the plexiglass. (Bundesarchiv)

(Above) The half-bowl shaped GSL-K 81Z turret rotated beneath the open end of the sloping rear gunner's canopy, its armor plating being integrally mounted in the turret. To avoid damage to the twin MG 81Z and its ammunition links, the guns could be held in position when the rear canopy was slid open — unlike the arrangement of the Ju 87B/R where the MG 15 travelled with the gun mount. The MG 81Z also featured VE 22 and later VE 42 twin sights. (Bundesarchiv)

Cockpit and Canopy Details

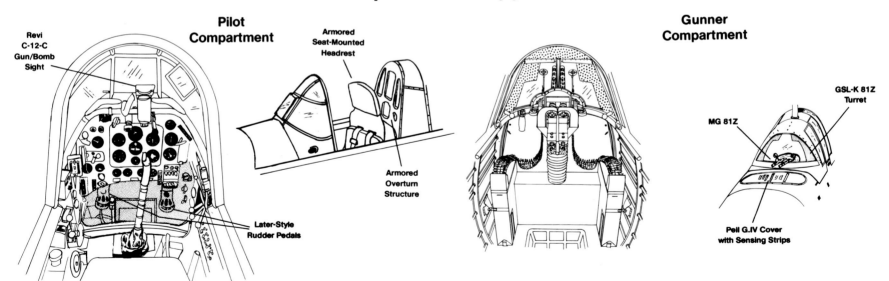

Pilot Compartment

Revi C-12-C Gun/Bomb Sight

Armored Seat-Mounted Headrest

Armored Overturn Structure

Later-Style Rudder Pedals

Gunner Compartment

GSL-K 81Z Turret

MG 81Z

Peil G.IV Cover with Sensing Strips

Ju 87D-2

Due to the success of the Ju 87B and R series as glider-tugs, the same attachments were adopted as a short-term production standard on a number of late production Ju 87D-1s under the designation Ju 87D-2. These aircraft also featured internal strengthening of the rear fuselage and tailwheel to enable them to pull heavier loads. While the Ju 87D-2 was specifically designated as a glider-towing *Dora* variant, practically any Ju 87 could be fitted with the glider towing hardware when necessary.

Ju 87D-3

By 1942 the *Stuka* was being increasingly used in the low-level ground assault role. To better meet these additional duties the Ju 87D-3 incorporated additional internal armor protection below the lower engine, radiator system and ventral fuselage. Since Allied troops had begun to 'steel their nerves' against the terror of the wind-driven sirens, they were less frequently fitted to the Ju 87D-3.

The first Ju 87D-3 entered service in May of 1942, and after initially sharing production lines with the Ju 87D-1, it became the standard *Stuka* variant by the close of the year. Initial production D-3s were externally identical to the D-1, and when fitted with a glider-towing attachment it was identical to the D-2 as well. However, running production line improvements provided the Ju 87D-3 with recognizable details such as fully exposed forward exhausts, simplified metal-strip wing walks (replacing the non-skid variety), and the total deletion of the now useless undercarriage siren mounting.

Production of the Ju 87D-3 continued through mid-1943, by which time it was being operated almost exclusively against the Soviet Union, having been among the last Ju 87 variants to see action in North Africa before the expulsion of Axis forces from that region in May of 1943. A total of 1,559 Ju 87D-3s are believed to have been produced, with 599 coming from Bremen-Lemwerder and 960 from Berlin-Tempelhof.

(Above) A pair of Ju 87D-1s of St.G 2 bank earthward, displaying their squadron codes on the undersides of their wings. Often, however, factory codes were maintained in this position to conserve time. The odd rudder markings on the background aircraft are typical of the additional vertical tail 'identifiers' which came into use toward the end of the war, particularly on stab (staff) flight aircraft. (Bundesarchiv)

(Right) To reduce accidents and cut down on the accumulation of mud and debris that built up on the inside of the undercarriage covers, the covers were frequently removed. The Ju 87D-1 in the foreground has had only its lower spats removed (which was common), while its squadron mate in the background has had its entire covers stripped away. (Bundesarchiv)

(Above) Even with its aerodynamic redesign, the 'Dora' was still an ominous-looking beast. The starboard supercharger intake now included a closable flap at its mouth. The new under-carriage legs were rather fragile looking in comparison to those of the earlier 'Berta', especially when considering the loads that the 'Dora' handled! (Bundesarchiv)

(Below) The groundcrew of 7./St.G 1 use a three wheeled bomb loader to attach an SC 250 bomb into position on the faired underwing bomb rack; the side-mounted ETC 50/VIII racks accommodated lighter loads. Light Blue (65) on the rear spat coverings was an effort to bet-ter 'sky camouflage' the Ju 87 when viewed from below — a practice which became more common on later nocturnal-assault Stukas. The placement of unit emblems on Luftwaffe vehicles was a common practice; in this instance on an Opel fuel truck. (Bundesarchiv)

Underwing Ordnance

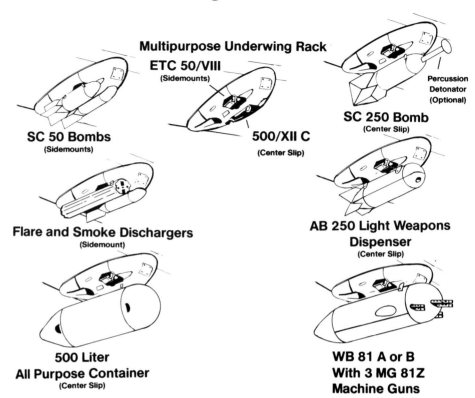

Multipurpose Underwing Rack
ETC 50/VIII
(Sidemounts)

SC 50 Bombs
(Sidemounts)

500/XII C
(Center Slip)

SC 250 Bomb
(Center Slip)

Percussion Detonator
(Optional)

Flare and Smoke Dischargers
(Sidemount)

AB 250 Light Weapons Dispenser
(Center Slip)

500 Liter
All Purpose Container
(Center Slip)

WB 81 A or B
With 3 MG 81Z
Machine Guns

(Below) When operating in the ground-attack role, WB 81A (seen here) or B weapons packs were carried to enhance strafing power. Armament was three MG 81Zs with 250 rounds-per-gun, mounted in trays within the streamlined canister. This arrangement was referred to as the 'watering can'. (Hans Redemann)

42

(Above) Oberst Ernst Kupfer of the re-formed II/St.G 2 fires up the engine of his Ju 87D-1 on the Eastern Front in 1942. The Gruppe's 'Knight of Bamberg' emblem carries a Kommandeur's chevron which was occasionally seen within Stuka units. The center section of the underwing bomb rack covers have been removed for easy access to the bomb rack mechanism — a common occurence. (Bundesarchiv)

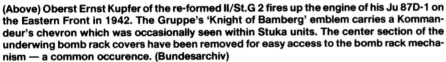

(Above Right) Ju 87D-1s of II/St.G 2 homeward bound, with their leather oleo shockbands ballooning in the slip stream. An unofficial variation of squadron coding was the use of the Staffel color on the fourth rather than the usual third fuselage letter, and sometimes both were painted in the staffel colors. The numbers on the wheel spats were usually co-ordinated with the aircraft letter (e.g. T6+AD would be 1, T6+BD would be 2, etc), however, this system was not strictly adhered to. (Bundesarchiv)

(Right) While the 'Dora's' redesign and the more powerful Junkers Jumo 211 J engine allowed for an optimum variety of bombloads, the usual bomb load remained one large underbelly weapon (either an SC 250 or SC 500) and two SC 50s on each under wing rack. Throughout the war a variety of percussion detonators were used, including some in-the-field improvisations. (Bundesarchiv)

(Above) Ju 87D-1s and D-3s were among the last Stukas to see service with St.G 3 in North Africa before the collapse of that front in the spring of 1943. By this time desert camouflage had come full circle, with many 'Doras' being thrown into battle still carrying their European colors. Due to rising Allied air-superiority, Stuka attrition in this region reached slaughterhouse proportions by the time the Ju 87D arrived. (Bundesarchiv)

(Above) 'Dora-3s' enroute to the fateful Battle of the Kursk salient, where the Stuka reached its eclipse as an effective dive-bomber and the Third Reich reached its eclipse in its last great armored offensive. With its days as a 'terror weapon' passed, the Ju 87 gradually dispensed with the propellor sirens, the mounting pods being totally eliminated on late production Ju 87D-3s as seen on the second aircraft. (Bundesarchiv)

Under Fuselage Bomb Rack

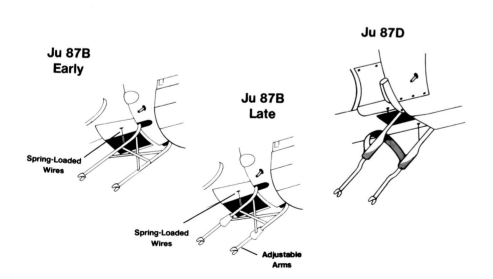

Ju 87B Early

Ju 87B Late

Ju 87D

Spring-Loaded Wires

Spring-Loaded Wires

Adjustable Arms

Horizontal Tail Supports

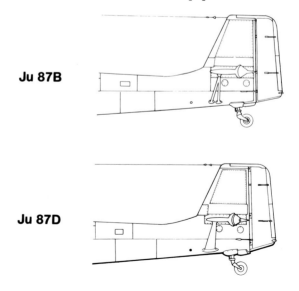

Ju 87B

Ju 87D

Exhaust

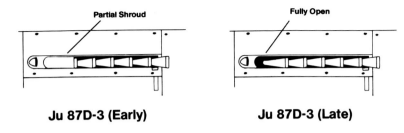

Partial Shroud

Fully Open

Ju 87D-3 (Early)

Ju 87D-3 (Late)

(Above) Rumanian Air Force personnel adjust the wingracks of a late Ju 87D-3 supplied to the Corpul 1 Aerian on the Eastern Front, during 1944. Ironically, some of these same Stukas were used against German forces during the Rumanian anti-Axis coup in September of 1944. Ju 87Ds were also used by Hungary, Bulgaria, Slovakia, and Italy.

Undercarriage

Siren Mount

Ju 87D-3 (Early)

Ju 87D-3 (Late)

(Below) The presence of a glider-towing attachment would indicate that this Dora is a D-2, however the lack of siren mounting pods and the fully exposed exhausts identifies it as a late Ju 87D-3 field-rigged as a tug. Precise identification of the Ju 87D-1, D-2 and D-3 aircraft often presented a challenge since components were interchangeable. (Bundesarchiv)

Wingwalk

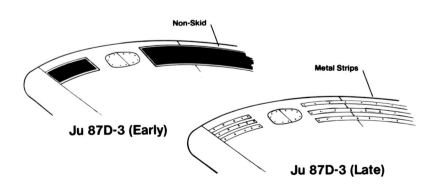

Non-Skid

Metal Strips

Ju 87D-3 (Early)

Ju 87D-3 (Late)

Ju 87D-3 Overwing Pods

The most unusual modification made to the *Stuka* was the addition of a large man-carrying container on each wing of a Ju 87D-3. These jettisonable, parachute-equipped containers were capable of carrying two tandemly-seated passengers and were neatly faired into the wing uppersurfaces with streamlining fillets. Windows were included in both sides of these 'personnel pods' providing 'see through' lateral vision for the pilot. Testing was conducted at the Research Institute — Graf Zeppelin at Stuttgart-Ruit during 1942. Had the pods proven successful, they would have been used for clandestine agent-dropping and nocturnal supply flights. However, problems were encountered, including proper release of the pods in flight, and the project was dropped.

Ju 87D-3 with Wing Pods

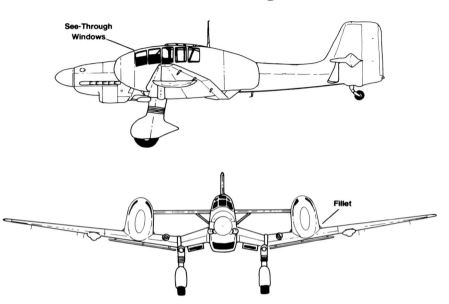

See-Through Windows

Fillet

Ju 87D-4 and Ju 87E

The Ju 87D-4 and the Ju 87E were each conceived during the *Dora's* 1941 developmental period, and while both shared common bonds, each was a separate proposal built to fulfill a specific combat role.

Ju 87D-4

The D-4, based on modified Ju 87D-1s and D-3s, featured an underbelly rack for mounting a 1,687 pound LT F5b torpedo or an interchangeable load of similar weight. Other than this, the D-4 was a standard *Dora* and was presumably to have served as a shore-based torpedo attack plane. That the torpedo carrying Ju 87D-4 failed to reach production status, was not surprising since Germany's torpedo-bombing needs were being met by faster twin-engine aircraft of greater load-lifting capacity such as the Heinkel He 111H-6 and the Junkers Ju 88A-17. Prototype D-4s were converted back to D-1 and D-3 standards.

Ju 87E

The Ju 87E program was devoted to the creation of a completely navalized Ju 87D with the same maritime gear and folding wings as the earlier Ju 87C project. Since plans for completing the *Graf Zeppelin* had not yet been abandoned, the conversion of the *Dora* as a more advanced ship-borne dive-bomber remained as a project. As an initial test-bed, an early Ju 87D-1, redesignated as a Ju 87D-1/to (torpedo), was converted for torpedo-carrying at the *Erprobungsstelle* Travemunde in late 1941 and tested at the *E-Stelle* at Rechlin. During the next year additional trials with other Ju 87 research prototypes (including Ju 87Cs) were conducted at Travemunde and at the rocket-research facility at Peenemunde-West; all committed to refining the catapult, arrester hook and rocket-assisted takeoff apparatus of the Ju 87E. However, by 1943 construction of the *Graf Zeppelin* was terminated and the tentative order for 115 Ju 87E-1s was cancelled.

Ju 87D Torpedo Mount

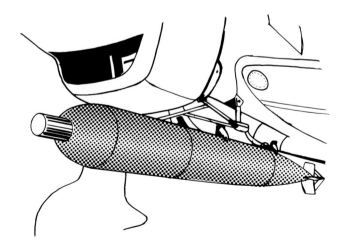

Ju 87D-5

In a final effort to refine the aging Ju 87 and increase its ground-attack capabilities, the Ju 87D-5 was introduced on the assembly lines in early 1943. While the fuselage was identical to that of the Ju 87D-3, the wingtips were tapered and lengthened to improve wing loading, increasing overall wingspan from 45 feet 3 1/4 inches to 49 feet 2 inches. For greater striking power, the wing-mounted MG 17s were replaced with a pair of Mauser MG 151 20MM cannons; the long barrels of which projected neatly from the center of the wing leading edges. Production line changes introduced on late Ju 87D-3s (fully exposed exhausts, metal strip wing walks and deletion of undercarriage siren mounts) were standardized on the Ju 87D-5. Additional changes to the Ju 87D-5 included modifications to the bomb-release system, lubrication system, jettisonable undercarriage and reinforcement of the pilot's ventral viewing window.

As the Ju 87D-5 became more committed to the ground-attack role, a number of late production machines were fitted with a number of optional modifications to the canopy apparently aimed at increasing pilot safety and comfort during these risky operations. These modifications included additional windshield braces with flat forward plexiglass panels, add-on sliding side windows, or a small plexiglass bubble atop the windscreen for relocation of the gunsight. These modifications were occasionally retro-fitted to earlier Ju 87D variants, particularly those serving in the nocturnal-assault capacity.

By the time the Ju 87D-5 entered service on the Eastern Front in July of 1943, the *Stuka* was breathing its last gasp as a dive-bomber. As the Germans went over to the defensive, fortified targets requiring pin-point attacks were replaced by masses of attacking Russian troops and armor, and growing Soviet airpower was rendering the Ju 87's vulnerable diving attacks extremely hazardous. The first unit to fly the Ju 87D-5 in combat was *III/St.G 2* during the critical armored offensive at Kursk, with the D-5 becoming commonplace with the *Stukagruppen* by summer's end. However, on 5 October 1943 dive-bombing was officially abandoned, with the *Stukageschwaders* being redesignated to *Schlachtgeschwaders* (ground-attack wings) and committed almost exclusively to the low-level ground attack role. Abbreviated unit designator was changed from *St.G* to *SG* — the D-5 seeing action with *SG 1, 2, 3, 5* and *77.*

The Ju 87D-5, and the *Stuka* in general, entered its twilight during late 1943 and early 1944 with the Ju 87 being increasingly replaced in the daylight attack role by the much by faster Focke-Wulf Fw 190 fighter-bomber. Only specialized Ju 87 units remained in service by the spring of 1945, occasionally retaining surplus stocks of surviving Ju 87D-5s on strength.

Of the original contract for 1,178 Ju 87D-5s, 771 are known to have been produced at Bremen-Lemwerder, with additional numbers rolling off the Tempelhof line before production tapered off during the summer of 1944. Thus the Ju 87D-5 became the last variant to see substantial quantity production, as well as the last variant to be manufactured specifically as a dive-bomber.

(Right) Ju 87D-5s believed to be on the assembly line at Bremen-Lemwerder. The portside position of the engine coolant header tank was introduced on the D-series, as was the underside oil-cooler assembly with port-offset shutters. The darkly primered panels on the forward fuselage are both standard and applique armor plate. (Bundesarchiv)

(Above) Major changes from the Ju 87D-3 to the Ju 87D-5 included the installation of long-barrelled 20MM MG 151 cannons and the more graceful long-span wing tips. While the harder-hitting D-5 was dangerously obsolete at the time of its introduction, it was still regarded as something of a stalwart battle-plane by its crews. These Ju 87D-5s of St.G 2 are enroute to the battle of Kursk, where the type first saw action. (Bundesarchiv)

47

(Right) After being driven from the skies as a dive-bomber, the Stukagruppen were consigned to the low-level ground-attack role under the redesignation Schlachtgeschwadern. The now useless underwing dive-brakes were frequently removed in the field usually leaving the mounting lugs hanging bare. A damaged undercarriage shock band was a common sight. (Bundesarchiv)

Ju 87D-3

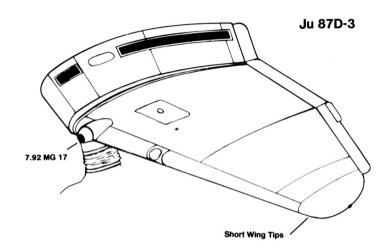

7.92 MG 17

Short Wing Tips

Ju 87D-5

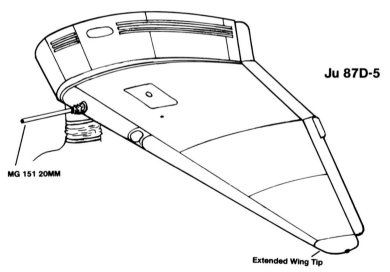

MG 151 20MM

Extended Wing Tip

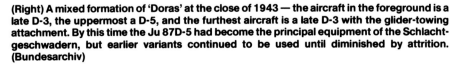

(Right) A mixed formation of 'Doras' at the close of 1943 — the aircraft in the foreground is a late D-3, the uppermost a D-5, and the furthest aircraft is a late D-3 with the glider-towing attachment. By this time the Ju 87D-5 had become the principal equipment of the Schlachtgeschwadern, but earlier variants continued to be used until diminished by attrition. (Bundesarchiv)

48

(Above) A squiggle-camouflaged Ju 87D-5 taxis into position during the winter of 1943-44, with SC 250s under its wings and an AB 500 anti-personnel weapons dispenser under its belly. What appear to be small underwing pylons directly in line with the barrels of the 20MM cannons are spent cartridge chutes, Ju 87D-5s being the first variant to have this feature. (Bundesarchiv)

(Above) Snow-camouflaged Ju 87D-5s with the nearly obligatory daylight escort of a pair of Messerschmitt Bf-109Gs. By the time the Ju 87D-5 entered service, many of the colorful unit emblems had virtually disappeared, except for theater markings and unit codes. (Bundesarchiv)

Late Canopy Options

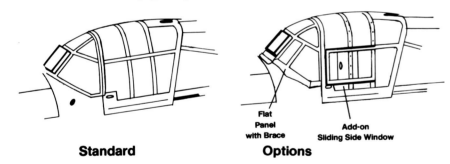

Standard   **Options**

Flat Panel with Brace

Add-on Sliding Side Window

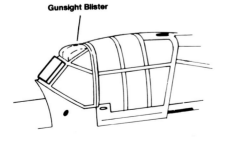

Gunsight Blister

Dive Brakes Development

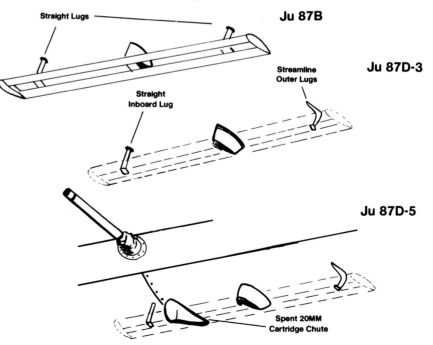

Straight Lugs   **Ju 87B**

Straight Inboard Lug

Streamline Outer Lugs   **Ju 87D-3**

Ju 87D-5

Spent 20MM Cartridge Chute

One of the Luftwaffe's most flamboyant and highly decorated airmen, Hans Ulrich Rudel racked up an unparalleled 2,530 sorties, the destruction of the Soviet battleship 'Marat', 519 tanks as well as numerous other enemy targets. Rudel finished the war with the rank of Oberst (Colonel) at the age of 28, and was still flying combat missions with one leg amputated. In this lineup of Ju 87D-5s of III/SG 2, Rudel's aircraft is foremost. By late 1943, the simplified crosses and Swastikas had become the norm for Ju 87's as well as the small low-visibility geschwader cipher of the fuselage code. The full code on Rudel's aircraft is T6+AD, with the 'A' and forward portion of the spinner in Light Green (25). The extreme tip of the spinner is believed to be in the Gruppe color of Yellow. (Bundesarchiv)

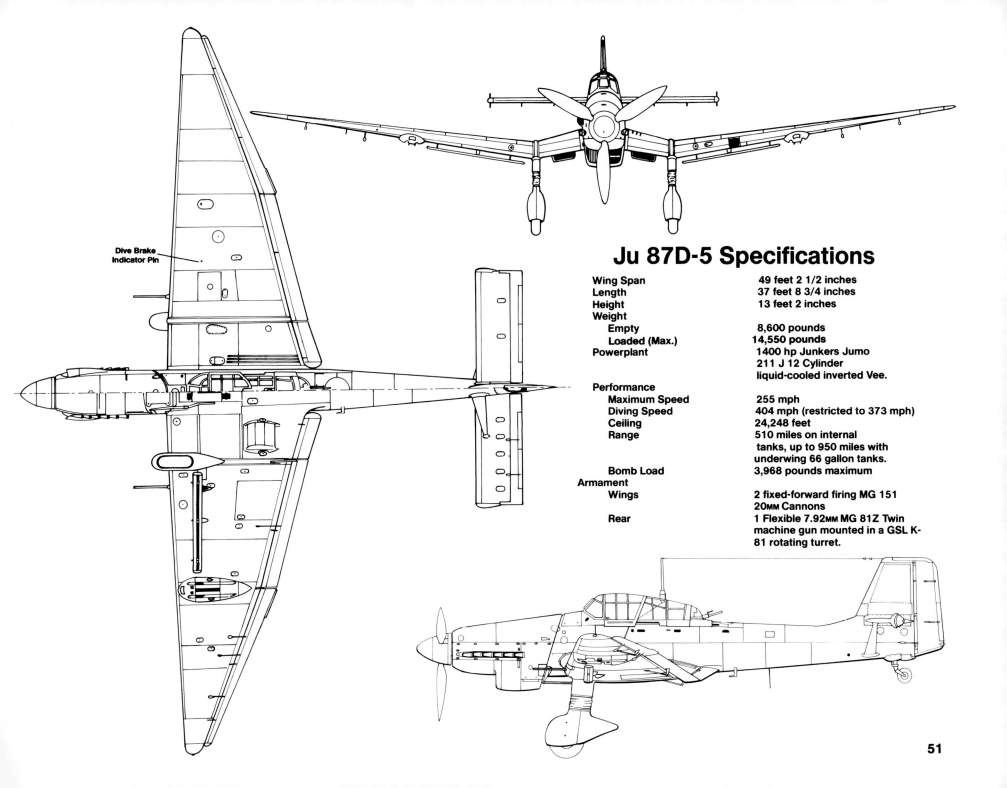

Ju 87D-5 Specifications

Dive Brake
Indicator Pin

Wing Span	49 feet 2 1/2 inches
Length	37 feet 8 3/4 inches
Height	13 feet 2 inches
Weight	
Empty	8,600 pounds
Loaded (Max.)	14,550 pounds
Powerplant	1400 hp Junkers Jumo 211 J 12 Cylinder liquid-cooled inverted Vee.
Performance	
Maximum Speed	255 mph
Diving Speed	404 mph (restricted to 373 mph)
Ceiling	24,248 feet
Range	510 miles on internal tanks, up to 950 miles with underwing 66 gallon tanks.
Bomb Load	3,968 pounds maximum
Armament	
Wings	2 fixed-forward firing MG 151 20ᴍᴍ Cannons
Rear	1 Flexible 7.92ᴍᴍ MG 81Z Twin machine gun mounted in a GSL K-81 rotating turret.

Ju 87D-7 and D-8

Early in the Russian campaign, the Red Air Force began carrying out nuisance raids at night against German positions using archaic aircraft such as the Polikarpov PO-2 biplane. In response, the *Luftwaffe* established several *Störkampfstaffeln* (harassment battle squadrons), equipping them with outdated types such as the Arado 66 and Gotha 145. During the reorganizational period in the fall of 1943, these somewhat 'freelance' units were consolidated into the more elite *Nachtschlachtgruppen* (night attack groups) and greater priority was given to providing them with more advanced aircraft. Since the Fw-190 was in the process of absorbing daytime ground-assault duties, strong well-armored Ju 87Ds were re-assigned to the nocturnal-assault role.

To meet the demands of night operations, some 300 Ju 87Ds were slated for recycling at the Hamburg-Harburg factory of Metallwerk Niedersachsen, Brinkmann and Mergell (Menibum) during late 1943 and 1944. Each aircraft was re-engined with the more powerful 1,500 hp Junkers Jumo 211 P engine, as well as adding flame-dampening tubes over the exhausts to prevent glare. These aircraft were usually stripped of their underwing dive-brakes, although the mountings often remained intact.

Modified short-span Ju 87D-3s were redesignated to Ju 87D-7's* and the long-span, cannon-armed Ju 87D-5 was redesignated to the Ju 87D-8. Both variants were subjected to optional modifications, resulting in a mixture of repositioned and reconfigured exhaust-tubes, flash-blinders over the muzzles of the wing guns, and the fitting of extra radio gear such as the FuG 16z and the more common FuG 25. Along with these radio additions, a number of later Ju 87D-8s also carried an improved D/F loop aerial which was mounted to a teardrop housing at the base of the canopy mast. The whip antenna for the FuG 25 was located, as usual, beneath the aft fuselage.

Ju 87D-7s and D-8s gradually re-equipped the *Nachschlacht* units throughout 1944, seeing action in Italy and along both the Eastern and Western Fronts. Usually operating at dusk or by moonlight, Ju 87s struck at moderate altitudes (3,000 to 7,000 feet) using a number of special tactics, such as flare-equipped marker/striker teams or by employing mobile ground-radio units to direct the *Stukas* against their targets (also known as the *Egon* procedure). Under the cover of darkness, losses to enemy groundfire were kept to a minimum, however, high accidental attrition became unavoidable due to increasingly ill-trained and inexperienced crews. Nocturnal *Stukas* further suffered from Allied nightfighter interceptions, but they continued operations until the downfall of the Third Reich, and were among the last Ju 87s to be encountered in substantial numbers by the advancing allies. *Nachtschlachtgruppen* known to be using the Ju 87 during the closing months of the war included *NSGr 1, 2, 4, 8, 9* and *10*.

---

* The designation of Ju 87D-6 had been reserved for a variant which did not see production.

(Above) Germany, 1945. Victorious G I's pose with a short-span Ju 87D-7 (converted D-3) which has had its wooden propeller blades blown off to prevent unauthorized use. The late-war modified windshield is detectable by its flat forward panels and extra braces. The 'spiral-schnauze' spinner was commonplace with aerial-assault units during the last year of the war.

(Right) Nachtschlacht Ju 87s occasionally featured night-camouflage over their uppersurface Green splinter — usually a squiggled or mottled pattern of Light Blue (76) or RLM Gray (02). 'Red G', believed to be operating with 2./NSG 9 in Italy, appears to have a squiggle pattern of RLM Gray over the upper surfaces. Another known scheme was the usual 70/71 splinter uppersurfaces with Black undersides. Styles of application varied considerably. (Bundesarchiv)

Ju 87D-7
(Converted Ju 87D-3)

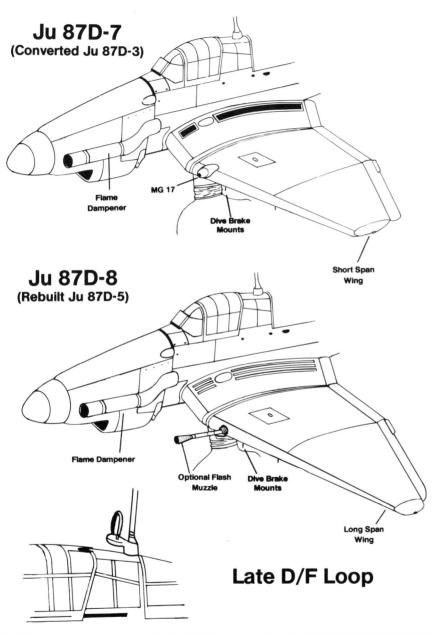

Flame Dampener

MG 17

Dive Brake Mounts

Short Span Wing

Ju 87D-8
(Rebuilt Ju 87D-5)

Flame Dampener

Optional Flash Muzzle

Dive Brake Mounts

Long Span Wing

Late D/F Loop

(Right) A groundcrewman inspects the MG 151 cannon bay of a Ju 87D-8 (converted D-5) in southern Italy. The optional flash-hinders have been fitted to the 20mm cannons and a bulbous extension has been added to the flame-dampening exhaust tubes to further reduce exhaust glare. The retention of dive-brakes on nocturnal-assault Stukas was not unheard of, but was somewhat unusual. (Bundesarchiv)

Ju 87G

Due to the rapid increase in Russian armor on the Eastern Front, the Ju 87 became one of many aircraft considered for conversion to the heavy-caliber tank-busting role. In December of 1942 a Ju 87D-1 was modified to carry two underwing mounted 600 pound Flak 18 37MM anti-tank guns. Satisfactorily tested at the Rechlin testing center, it served as the prototype for the Ju 87G series. During February of 1943, several Ju 87G-1s (converted from late-model Ju 87D-3s) were sent to the experimental anti-tank unit, *Panzerjägdkommando Weiss*, where they were successfully flown into combat by selected veteran pilots. As a result of their achievements, the short wing span Ju 87G-1 was ordered into limited conversion. During the months that followed, long-span Ju 87D-5 airframes were rebuilt as the main production standard under the designation of Ju 87G-2.

On both variants, the light wing mounted armament and bomb racks were removed. The two 37MM cannons (also known as the Bord Kannone 3.7CM) were suspended beneath the outer wings, encased in streamlined pods protecting the breech mechanism. Maintenance of the cannon was effected by hinged side panels and detachable endcaps. The ammunition trays for the three pound shells (twelve rounds per gun) passed through the housings which protruded on either side. With a muzzle velocity of 2,610 to 2,820 feet per second, the Wolfram-core armor-piercing rounds were effective against most Russian tanks, while slightly lighter high-explosive rounds could be used against softer targets.

Since the Ju 87G-1 was a *conversion* of existing Ju 87D-3 aircraft, the wing bulges of the MG 17 machine guns were still apparent and the braces for the discarded underwing dive-brakes were often left in place. However the Ju 87G-2 was a *rebuilt* Ju 87D-5 with clean wing leading edges and an absence of all extraneous mountings. Further detail differences could be found among operational machines which varied considerably, including flame-dampening tubes for twilight attacks and the occasional retention of the lighter fixed wing guns to serve as aiming guides with tracer ammunition. For the most part, however, Ju 87Gs were flown into action as-delivered.

The first Ju 87G's went into combat during the Kursk Offensive with *St.G 1* and St.G 2, and by late 1943 *Schlachtgeschwaders 3* and 77 were also equipped with the sparingly issued Ju 87G within specialized *10. Panzerjäger Staffeln*. While the Ju 87G was the least maneuverable of all Ju 87 variants, they were among the more dependable and efficient of the *Luftwaffe's* 'tank-busters'. Most were concentrated on the Eastern Front, but during the closing months of the war, they were employed on the Western Front as well.

Out of a final production order for 208 rebuilt Ju 87G-2s, 174 were produced at Bremen-Lemwerder before all Ju 87 manufacturing ceased in October of 1944. With these last examples, the sadly outdated Stuka had managed to remain in production for just under eight years.

Ju 87G-1 Bord Kannone 3.7 CM

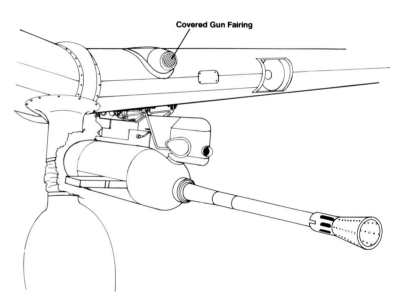

Covered Gun Fairing

(Left) The premier 'tankbuster' Stuka to see combat was the Ju 87G-1, a conversion of late model Ju 87D-3 airframes, with the installation of a 37MM Flak 18 cannon under each wing. Thanks to its destructive power and distinctive six foot Flak 18 cannon barrels, the 'Gustav' was christened with such nicknames as 'Kanonenvogel' (Cannonbird), 'Panzerknacker' (Tank cracker) or, more simply the 'Stuka mit den Langen Stangen' (Stuka with the long rods). It lived up to its knicknames. (Bundesarchiv)

(Above) One of the first Ju 87G-1s to join 'Panzerjagdkommando Weiss' is carrying factory codes, Yellow Russian front markings, and a Red tipped spinner. It is believed that this 'Gustav' was one of the earlier mounts of Hans Ulrich Rudel, who is also believed to have suggested the 'T-34' emblem on the cowling. As with most short-span Ju 87G-1s, the pivoting lugs for the discarded dive-brakes are still in place under the wings, and the MG17 machine gun ports have been faired over. (Bundesarchiv)

(Above) Six-shot clips of armor-penetrating Wolfram-core 37mm ammunition were loaded into trays through a hinged panel; the rounds were either armor-piercing or blunt-tipped for soft-skinned targets. The structure of the underwing cannons and pods were identical and interchangeable, with the loading aperture and external connections being offset to starboard on each. (Bundesarchiv)

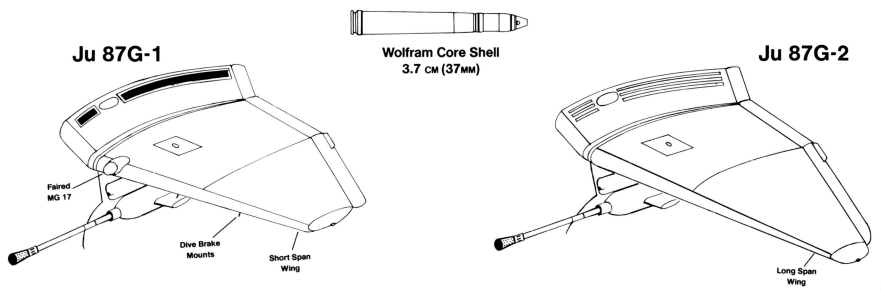

Ju 87G-1

Wolfram Core Shell
3.7 cm (37mm)

Ju 87G-2

Faired MG 17

Dive Brake Mounts

Short Span Wing

Long Span Wing

(Above) An armorer covers his ears against the pop of the Flak 18 3.7 Bordkanone during field calibration. The Flak 18 was capable of disabling just about any Russian tank, and when employed in stern attacks against the thinner armor plating of engine compartments, the results were devastating. (Bundesarchiv)

(Above Right) Like its predecessors, the 'Gustav' received its share of winter whitewash. Typical of Ju 87G-1s was the use of side applique armor, which, more often than not, was absent on production Ju 87G-2s. A number of sources have mentioned that the underwing cannons could be replaced by bomb racks when necessary, but little has surfaced to support this belief, and is highly unlikely since Tankbusters were so desperately needed against the waves of Soviet armor. (Bundesarchiv)

(Right) Produced in greater numbers than the Ju 87G-1, the re-built Ju 87G-2 was based on the long wing span Ju 87D-5 and was a much cleaner aircraft; all traces of the underwing dive-brake mounts and light caliber wing guns being deleted. For access to the cannon mechanism, the side panels of the cannon pods were hinged to fold downward and the end caps were removable fore and aft. Cannon installations were the same for both the G-1 and G-2. (Hans Obert)

(Above) An abandoned Ju 87G-2 shares a hangar with a cannibalized Ju87D-5, the 'Gustav' apparently having suffered light damage below the pilot's windscreen. Scattered groups of ground-attack and tankbuster Stukas vainly harassed the advancing Allies in the East and West during the last days of the war. On the 'Dora' at right can be seen the remnants of a Yellow undercowling, which was a wide spread variation of late war tactical markings, and was also to be found on some 'Gustavs'. (Richard F. Grant)

(Above) War's end, 8 May 1945. Luftwaffe personnel strike a different kind of propaganda pose after landing their Ju 87G-2 at a forward Allied air base. As countless German aircraft 'airlifted' their crews to the comparative safety of the West, the anachronistic Stuka made its swan song. All armament, including the barrels of the underwing Flak 18s, have been stripped as a gesture of passivity. (US Air Force)

Ju 87H-1

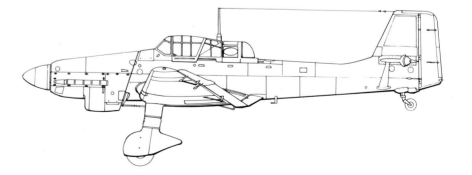

Ju 87H Trainers

During the latter half of the war the tremendous losses of veteran pilots was particularly hard felt by the *Stukagruppen*. In order to increase the rapidity of training fresh *Stuka* crews, as well as the conversion training of airmen having experience on other aircraft, a limited number of Ju 87D-1, D-3, D-5, D-7 and D-8s were modified into Ju 87H-1 through H-8 trainers during the course of 1944. Externally, these aircraft were identical to their *Dora* counterparts, but were stripped of armament, bomb racks and the rear GSL-K 81 turrets. Complete dual controls and seating from Arado 96 trainers were fitted into the rear cockpit, and transparent blisters were added to the aft canopy to improve the instructor's forward vision.

Beyond their intended use as trainers, some Ju 87Hs are believed to have been futilely rigged for combat at the close of hostilities in 1945, but details of these conversions have remained understandably sketchy. Although many war-weary and outdated Ju 87s had served as trainers prior to the Ju 87H, it was the only variant specifically designated and equipped for the training role.

Luftwaffe Aircraft

From
squadron/signal publications

FOCKE WULF **FW 190** IN ACTION

squadron/signal publications
AIRCRAFT NO. NINETEEN

Messerschmitt Bf 109 in action Part 1

squadron/signal publications
AIRCRAFT No. 44

Messerschmitt Bf 109 in action Part 2

SPECIAL 8 EXTRA PAGES

Aircraft Number 57
squadron/signal publications

Junkers Ju 88 in action Part 1

squadron/signal publications, inc.   Aircraft Number 85

SPECIAL 8 EXTRA PAGES

Junkers Ju 88 in action Part 2

Aircraft Number 113
squadron/signal publications

SPECIAL 8 EXTRA PAGES

FOCKE-WULF Fw 189 in action

Aircraft Number 142
squadron/signal publications